Beast Academy

By Art of Problem Solving

MATH PRACTICE 4C

JASON BATTERSON
SHANNON ROGERS

© 2014, 2016, 2017 AoPS Incorporated. All Rights Reserved.

Reproduction of any portion of this book without the written permission of AoPS Incorporated is strictly prohibited, except for fair use or other noncommercial uses as defined in sections 107 and 108 of the U.S. Copyright Act.

Published by: AoPS Incorporated
　　　　　　　　15330 Avenue of Science
　　　　　　　　San Diego, CA 92128
　　　　　　　　info@BeastAcademy.com

ISBN: 978-1-934124-55-0

Written by Jason Batterson, Shannon Rogers, and Deven Ware
Book Design by Lisa T. Phan
Illustrations by Erich Owen
Grayscales by Greta Selman

Visit the Beast Academy website at BeastAcademy.com.
Visit the Art of Problem Solving website at artofproblemsolving.com.
Printed in the United States of America.
2022 Printing.

Contents:

How to Use This Book	4
Chapter 7: Factors	6
Chapter 8: Fractions	38
Chapter 9: Integers	72
Hints for Selected Problems	104
Solutions	108

This is Practice Book 4C in the Beast Academy level 4 series.

For more resources and information, visit BeastAcademy.com/resources.

INTRODUCTION: How to Use This Book

"This is Beast Academy Practice Book 4C."

"Each chapter of this Practice book corresponds to a chapter from Beast Academy Guide 4C."

"The first page of each chapter includes a recommended sequence for the Guide and Practice books."

"You may also read the entire chapter in the Guide before beginning the Practice chapter."

Use this Practice book with Guide 4C from BeastAcademy.com.

Recommended Sequence:

Book	Pages
Guide:	14–31
Practice:	7–14
Guide:	32–47
Practice:	15–37

You may also read the entire chapter in the Guide before beginning the Practice chapter.

CHAPTER 7
Factors

Use this Practice book with Guide 4C from BeastAcademy.com.

Recommended Sequence:

Book	Pages:
Guide:	14-31
Practice:	7-14
Guide:	32-47
Practice:	15-37

You may also read the entire chapter in the Guide before beginning the Practice chapter.

Factors of a number n are the numbers that n is divisible by.

EXAMPLE | Which of the numbers below are factors of 6?

1   2   3   4   5   6

6 is divisible by 1, 2, 3, and 6.
6 is not divisible by 4 or 5.

Therefore, **1, 2, 3, and 6** are factors of 6.

Introduction — FACTORS

PRACTICE | Answer each question below.

1. Circle all the numbers below that are factors of 18.

   1   2   3   4   5   6   7

2. Circle all the numbers below that are factors of 21.

   1   3   5   6   7   14   42

3. Circle all the numbers below that are factors of 72.

   2   5   7   9   24   36   72

4. Circle all the numbers below that have 5 as a factor.

   12   15   21   30   38   45   53

5. Circle all the numbers below that have 3 as a factor.

   2   6   16   22   36   45   65

6. Circle all the numbers below that have 8 as a factor.

   4   12   16   22   32   48   96

7. Which two numbers are factors of both 45 and 21?   7. _____ and _____

FACTOR PAIRS

EXAMPLE | How many factors does 48 have?

We list pairs of numbers that can be multiplied to get 48. We begin by looking at the pairs with the smallest factors and organize our work by writing the smaller number in each pair first.

<u>**48:**</u>
1 × 48
2 × 24
3 × 16
4 × 12
6 × 8

Then, since 7 × 7 = 49, any number that is **larger** than 7 has to be multiplied by a number that is **smaller** than 7 to get 48.

We have found all the factors of 48 that are smaller than 7, so every factor of 48 that is larger than 7 is paired with a factor that is smaller than 7. There are no other factors of 48.

All together, we count **10** factors of 48.

> Writing a number's factors in pairs helps us stay organized, which makes finding all the factors easier!

PRACTICE | List the factor pairs for each number given below.

8. List each of the four factor pairs whose product is 30.

<u>**30:**</u>
____ × ____
____ × ____
____ × ____
____ × ____

9. List each of the four factor pairs whose product is 56.

<u>**56:**</u>
____ × ____
____ × ____
____ × ____
____ × ____

10. List each of the factor pairs whose product is 35.

<u>**35:**</u>

11. List each of the factor pairs whose product is 54.

<u>**54:**</u>

PRACTICE | Answer each question below.

12. List each of the factor pairs whose product is 26.

<u>**26:**</u>

13. List each of the factor pairs whose product is 40.

<u>**40:**</u>

14. How many of the factors of 42 are odd?

14. _____

15. List all of the factors of 48 that are also factors of 30.

15. _____

16. Grogg draws a rectangle that has side lengths that are whole numbers of inches. The rectangle's area is 52 square inches. How many different heights are possible for Grogg's rectangle?

16. _____

17. ★ ★ Winnie picks a number n. When Winnie divides 59 by n, she gets a remainder of 3. List all possible values of n.

17. _____

FACTORS
Factor Pairs

PRACTICE | Answer each question below.

18. List each of the factor pairs whose product is 100.

100:

19. List each of the factor pairs whose product is 64.

64:

20. How many *different* numbers are factors of 100?

20. _____

21. How many *different* numbers are factors of 64?

21. _____

22. Explain why every perfect square has an odd number of factors.
Challenge: Explain why perfect squares are the **only** numbers with an odd number of factors.

23. There are four numbers less than 50 that have *exactly* three factors. The smallest of these numbers is 4. What are the other three?

23. _____, _____, and _____

24. Elliott has a room with 100 light switches, numbered from 1 to 100. Each light switch is turned off. Elliott invites 100 friends over and gives each friend a different number from 1 to 100. Friend number 1 flips every switch that is a multiple of 1 (all of them). Friend number 2 then flips every switch that is a multiple of 2. Friend number 3 flips every multiple of 3. This continues until all 100 friends have flipped switches, with friend number 100 flipping only the 100th switch. After all of this flipping, which light switches are on?

Note: When a friend **flips** a light switch, if it's on, they turn it off. If it's off, they turn it on.

10 | Beast Academy Practice 4C

EXAMPLE | Which of the numbers below are prime?

21, 22, 23, 24, 25

> A number with exactly two factors, 1 and itself, is called a *prime number.*
>
> Numbers that are prime are often simply called *"primes."*

We list the factor pairs of each of these five numbers:

21:	**22:**	**23:**	**24:**	**25:**
1×21	1×22	1×23	1×24	1×25
3×7	2×11		2×12	5×5
			3×8	
			4×6	

Only 23 has exactly two factors (1 and 23). So, **23** is the only prime number in the group.

21, 22, 24, and 25 each have more than two factors, so each of these four numbers is **composite**.

> A positive number with more than two factors is called a *composite* number.
>
> 0 and 1 are neither prime nor composite.

PRACTICE | Answer each question below.

25. Is 39 *prime* or *composite*?　　　　　　　　　　　　　　　　　25. _____

26. Is 47 *prime* or *composite*?　　　　　　　　　　　　　　　　　26. _____

27. How many even numbers are prime?　　　　　　　　　　　　27. _____

28. Circle all of the prime numbers in the list below.

　　　40　　　41　　　57　　　63　　　65　　　79　　　1,728

29. Circle all of the composite numbers in the list below.

　　　29　　　37　　　42　　　49　　　51　　　59　　　395

30. Exactly nine primes are less than 25. List them in order from least to greatest.

　　____, ____, ____, ____, ____, ____, ____, ____, and ____

Beast Academy Practice 4C

Sieve of Eratosthenes — FACTORS

PRACTICE — Complete the activity below to answer the questions that follow.

31. Start with the grid of numbers below and perform the following process to identify all of the primes less than 100:
    1. Circle the smallest number that is not already circled or crossed out.
    2. Cross out all of the multiples of the number you circled in Step 1, except the circled number itself.
    3. Repeat Steps 1 and 2 until every number on the grid is either circled or crossed out.

	2	3	4	5	6	7	8	9	10
11	12	13	14	15	16	17	18	19	20
21	22	23	24	25	26	27	28	29	30
31	32	33	34	35	36	37	38	39	40
41	42	43	44	45	46	47	48	49	50
51	52	53	54	55	56	57	58	59	60
61	62	63	64	65	66	67	68	69	70
71	72	73	74	75	76	77	78	79	80
81	82	83	84	85	86	87	88	89	90
91	92	93	94	95	96	97	98	99	100

This algorithm is called the **Sieve of Eratosthenes**. You can use these same steps to identify the primes between 2 and *any* number!

You can print out more copies of this grid at BeastAcademy.com.

32. How many primes are less than 100?
    32. _____

33. What is the largest two-digit prime?
    33. _____

34. What is the most common units digit among the two-digit primes?
    34. _____

35. ★ Lizzie subtracts one prime from another and gets 21. What will Lizzie get when she adds the same two primes?
    35. _____

FACTORS

Sieve of Eratosthenes

Math beasts have given names to many different types of prime numbers.

You don't need to memorize these names.

36. **Twin primes** are pairs of prime numbers that differ by 2. For example, 41 and 43 are twin primes. How many pairs of two-digit twin primes are there?

36. _____

37. An **emirp** is another special type of prime. When the digits of an emirp are written in reverse order, the result is a **different** prime number. For example, 13 and 31 are emirps. List all the pairs of two-digit emirps.

37. _____

38. A **Sophie Germain prime** is a prime p such that $(2 \times p) + 1$ is also prime. For example, 3 is a Sophie Germain prime because $(2 \times 3) + 1 = 7$ is also prime. List all the Sophie Germain primes that are less than 50.

38. _____

39. An **Eisenstein prime** is a prime number that is one less than a multiple of 3. For example, 71 is an Eisenstein prime because $71 = 72 - 1 = (3 \times 24) - 1$. List all the Eisenstein primes that are less than 100.

39. _____

40. A **Mersenne prime** is a prime number that is one less than a power of 2. For example, 7 is a Mersenne prime because $7 = 8 - 1 = 2^3 - 1$. List all the Mersenne primes that are less than 100.

40. _____

Fun fact: At the time of this book's printing, the largest known prime is a Mersenne prime equal to $2^{74,207,281} - 1$. It was found in 2016 and has more than **22 million** digits! Math beasts continue to look for larger primes.

Just like there is no largest number, there is no largest prime!

FACTORS — Divisibility Review

EXAMPLE Which of the numbers below are divisible by 8?

888   4,850   16,864

We can write 888 as the sum of three multiples of 8. So, **888 is divisible by 8**.

$$888 = 800 + 80 + 8$$
$$= (8 \times 100) + (8 \times 10) + (8 \times 1)$$
$$= 8 \times (100 + 10 + 1)$$
$$= 8 \times 111.$$

We can write 4,850 as 4,800+48+2. Since 4,800 and 48 are both multiples of 8, we know that 4,800+48 is a multiple of 8. Therefore, 4,850 is 2 more than a multiple of 8.
So, **4,850 is not divisible by 8.**

$$4,850 = 4,800 + 48 + 2.$$
$$= (8 \times 600) + (8 \times 6) + 2$$
$$= 8 \times (600 + 6) + 2$$
$$= 8 \times 606 + 2.$$

We can write 16,864 as the sum of three multiples of 8.
So, **16,864 is divisible by 8**.

$$16,864 = 16,000 + 800 + 64$$
$$= (8 \times 2,000) + (8 \times 100) + (8 \times 8)$$
$$= 8 \times (2,000 + 100 + 8)$$
$$= 8 \times 2,108.$$

When we add multiples of a number n, the result is also a multiple of n.

PRACTICE Answer each question below.

41. Circle every number below that is divisible by 7.

7,007   49,049   147,285   700,721   35,035,353

42. Circle every number below that is divisible by 8.

4,828   8,064   80,402   484,560   484,024

43. Circle every number below that is divisible by 9.

9,045   3,610   45,726   63,027   365,418

EXAMPLE | Without computing the quotient, what is the remainder when 30,000 is divided by 9?

30,000 = 10,000+10,000+10,000.

Since 10,000 = 9,999+1 = (9×1,111)+1, 10,000 is one more than a multiple of 9.

30,000 = 10,000+10,000+10,000
 = (9,999+1)+(9,999+1)+(9,999+1)
 = (9,999+9,999+9,999)+1+1+1
 = (9,999+9,999+9,999)+3.

Since 9,999 is divisible by 9, (9,999+9,999+9,999) is divisible by 9. 30,000 is 3 more than (9,999+9,999+9,999).

So, 30,000÷9 has remainder **3**.

PRACTICE | *Without computing a quotient*, give the remainder when each number or sum is divided by 9.

44. 99,999+999   44. _____

45. 100   45. _____

46. 10+10+10+10+10+10+10+10   46. _____

47. 1,000+1,000+1,000+1,000   47. _____

48. 10,000+1,000+100   48. _____

49. 5,000   49. _____

50. 3,000+30   50. _____

51. 10,000+2,000+300   51. _____

52. 50,000+4,000+300+20+1   52. _____

Beast Academy Practice 4C

FACTORS: Divisibility by 9 and 3

EXAMPLE | Is 14,382 divisible by 9?

If the sum of a number's digits is divisible by 9, then the number is divisible by 9.
If the sum of a number's digits is not divisible by 9, then the number is not divisible by 9.

The sum of the digits of 14,382 is
1+4+3+8+2 = 18.

Since 18 is divisible by 9, so is 14,382.

Therefore, **14,382 is divisible by 9**.

Review pages 32-39 in the Guide to see why these divisibility tests work!

EXAMPLE | Is 20,200 divisible by 3?

If the sum of a number's digits is divisible by 3, then the number is divisible by 3.
If the sum of a number's digits is not divisible by 3, then the number is not divisible by 3.

The sum of the digits of 20,200 is 2+0+2+0+0 = 4.
Since 4 is not divisible by 3, neither is 20,200.

Therefore, **20,200 is not divisible by 3**.

PRACTICE | Use the divisibility tests for 3 and 9 to help you determine which numbers are divisible by 3 and 9.

53. Circle every number below that is divisible by 3.

702   1,689   8,213   14,695   198,664   594,231

54. Circle every number below that is divisible by 9.

504   3,152   2,853   16,572   374,184   482,527

PRACTICE | Answer each question below.

55. $\boxed{4\,|\,1\,|\,A\,|\,6}$ is a four-digit number that is divisible by 9. Which digit is A?

55. $A =$ _____

56. List all of the digits that could replace B to make $\boxed{2\,|\,0\,|\,B\,|\,0}$ a four-digit number that is divisible by 3.

56. _____

57. ★ How many three-digit multiples of 9 can be made by arranging three of the four digits below?

$\boxed{1}\;\boxed{3}\;\boxed{5}\;\boxed{7}$

57. _____

58. ★ Grogg wants to write a multiple of 9 that uses only 1's and 2's as digits. What is the smallest multiple of 9 that Grogg can write using only 1's and 2's?

58. _____

59. ★ How many three-digit multiples of 3 can be made by arranging three of the four digits below?

$\boxed{2}\;\boxed{4}\;\boxed{6}\;\boxed{8}$

59. _____

60. ★★ S, T, and U are different digits. Each of the three-digit numbers $\boxed{S\,|\,T\,|\,U}$, $\boxed{T\,|\,U\,|\,S}$, and $\boxed{U\,|\,S\,|\,T}$ is even, and each is divisible by 9. What is $S+T+U$?

60. _____

Divisibility Tests

You can review why the tests for 2, 4, 5, 10, and 25 work in Guide 4B!

Review why the tests for 3 and 9 work on pages 32-39 of Guide 4C.

Number	Divisibility Test	Examples
2	Units digit is even	40<u>8</u> and 60<u>2</u>
3	Sum of digits is a multiple of 3	564 (5+6+4 = 15)
4	Number formed by last two digits is divisible by 4 (Ends in 00, 04, 08, ..., 88, 92, or 96)	6<u>08</u> and 8,9<u>60</u>
5	Units digit is 0 or 5	98<u>5</u> and 67<u>0</u>
9	Sum of digits is a multiple of 9	675 (6+7+5 = 18)
10	Units digit is 0	74<u>0</u> and 20,86<u>0</u>
25	Number formed by last two digits is divisible by 25 (Ends in 00, 25, 50, or 75)	2<u>75</u> and 12,6<u>25</u>
100	Ends in 00	9,2<u>00</u> and 12,6<u>00</u>

For one-digit numbers, we include leading zeros where necessary to use these tests.
For example, 0 = 00 and 8 = 08 are divisible by 4.

If a number does **not** satisfy the divisibility test for n, that number is **not** divisible by n.
For example, a number that does not have units digit 0 or 5 is not divisible by 5.

PRACTICE | Answer each question below.

61. Fill in the blanks in the number below to make a four-digit number that is divisible by both 9 and 25.

$$4\ \underline{}\ 7\ \underline{}$$

62. Fill in the blanks in the number below to make a four-digit number that is divisible by 4, by 5, and by 9.

$$5\ 7\ \underline{}\ \underline{}$$

63. ★ In how many ways can you arrange the four digits below to make a four-digit number that is even and divisible by 3?

$$\boxed{2}\ \boxed{3}\ \boxed{4}\ \boxed{9}$$

63. _____

64. ★ What is the smallest multiple of 3 that uses only the digits 8 and 9, with at least one of each?

64. _____

PRACTICE | Name the smallest prime factor of each number.

65. 736: _____

66. 1,625: _____

67. 217: _____

68. 81,135: _____

Prime Factorization

EXAMPLE What is the prime factorization of 60?

Every composite number can be written as the product of prime numbers.

We create a factor tree to find all of the prime factors of 60. We begin by factoring 60 into 6×10.

We call this product of primes the **prime factorization** of the composite number.

Next, we factor 6 into 2×3 and 10 into 2×5. We circle each of the prime factors at the bottom of the tree.

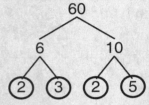

The prime factorization of a prime number is just the number itself.

Since there are no composite numbers left to factor, we are finished! The prime factorization of 60 is 2×3×2×5.

We generally order the factors from least to greatest: **2×2×3×5**. We usually write prime factorizations using exponents: **$2^2 \times 3 \times 5$**.

We check that the product equals 60:
$2^2 \times 3 \times 5 = 2 \times 2 \times 3 \times 5 = 60$.

Note that we could have begun by factoring 60 into 2×30, 3×20, 4×15, 5×12, or 6×10. No matter how we begin our factor tree, we will always end up with the same prime factorization!

PRACTICE Fill in the missing numbers in each factor tree below to determine the prime factorization of each number.

69.

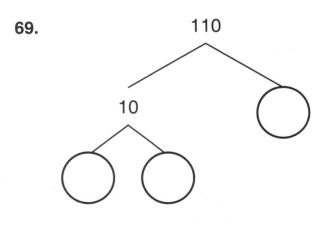

110 = _____

70.

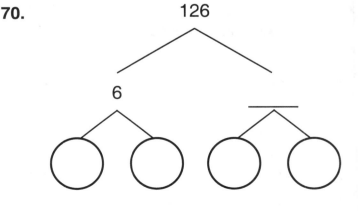

126 = _____

PRACTICE | Draw a factor tree to help you determine the prime factorization of each number below. Order the primes from least to greatest and use exponents for repeated factors, as in the example on the previous page.

71. 140 = _____

72. 72 = _____

73. 196 = _____

74. 465 = _____

FACTORS
Prime Factorization

PRACTICE | Draw a factor tree to help you determine the prime factorization of each number below. Order the primes from least to greatest and use exponents for repeated factors.

75. 600 = _____

76. 525 = _____

PRACTICE | Use the prime factorizations you found above to help you determine the prime factorization of each number below.

77. 1,800 = _____

78. 1,050 = _____

79. 6,000 = _____

80. 5,250 = _____

81. 300 = _____

82. 105 = _____

| EXAMPLE | What is the prime factorization of 127?

FACTORS
Prime Factorization

We use division and our divisibility tests to look for factors of 127.

7 is not an even digit, so 127 is not divisible by 2.

1+2+7 = 10 is not a multiple of 3, so 127 is not divisible by 3.

Since 4 = 2×2, every number that has 4 as a factor also has 2 as a factor. 127 is not divisible by 2, so 127 is not divisible by 4.

Similarly, every composite number has at least one prime factor. So, we only need to check for **prime** factors!

127 does not end in 0 or 5, so 127 is not divisible by 5.

127÷7 has remainder 1, so 127 is not divisible by 7.

127÷11 has remainder 6, so 127 is not divisible by 11.

127÷13 has remainder 10, so 127 is not divisible by 13.

Then, since 13×13 = 169, any number that is **larger** than 13 has to be multiplied by a number that is **smaller** than 13 to get 127.

So, we don't need to check any more primes. The only factors of 127 are 1 and 127.

Therefore, 127 is prime, and the prime factorization of 127 is just **127**.

Review why these are the only numbers we need to test on pages 40-47 of the Guide!

| **PRACTICE** | Write the prime factorization of each number on the line that follows. Order the primes from least to greatest, and use exponents for repeated factors.

83. 87 = _____

84. 113 = _____

85. 441 = _____

86. 910 = _____

87. 406 = _____

88. 357 = _____

Factor Blobs

In a **Factor Blob** puzzle, we circle "blobs" of two or more factors whose product is a given target number.

Every square in a blob must share at least one edge with another square in the blob. Blobs may not overlap. The goal is to use every number in the grid in a blob.

Target: 30

6	3	5
5	2	3
2	15	10

EXAMPLE | Solve the Factor Blob puzzle on the right.

Since 6×5 = 30, the 6 in the top left corner must be paired with a 5. So, we pair the 6 with the 5 below it in a blob. Similarly, since 10×3 = 30, the 10 in the bottom right corner must be paired with the 3 above it.

We then group the remaining numbers into blobs as shown so that the product of the numbers in each blob is 30.

In some cases, it may be useful to find the prime factorization of the target number and of the composite numbers in the grid.

PRACTICE | Solve each Factor Blob puzzle below.

89. Target: **45**

9	5	15
3	5	3
3	15	3

90. Target: **28**

7	1	1
14	2	2
4	7	2

91. Target: **750**

25	15	5
30	10	2
3	5	25

92. Target: **63**

3	7	3
21	7	7
3	9	9

FACTORS
Factor Blobs

PRACTICE | Solve each Factor Blob puzzle below.

93. Target: **42**

7	7	2	2
3	7	3	7
2	3	2	3

94. Target: **60**

6	5	3	5
2	2	2	2
6	10	3	10

95. Target: **140**

5	5	7	1
2	2	7	1
2	2	1	1

96. Target: **24**

2	4	4	1
2	2	3	1
3	2	3	2

97. ★ Target: **900**

25	4	9	10
2	15	6	3
75	12	5	30

98. ★ Target: **99**

3	3	9	11
33	3	11	3
3	1	1	11

99. ★ Target: **96**

24	4	6	16
4	6	1	6
6	4	16	16

100. ★ Target: **72**

6	6	6	6
6	2	18	2
6	2	2	2

Pyramid Descent

In a **Pyramid Descent** puzzle, the goal is to find a path of touching squares, one per row, from the top to the bottom of the pyramid.

The product of the numbers on the path must equal the number shown above the pyramid.

EXAMPLE Complete the Pyramid Descent puzzle to the right.

140

```
        1
      2   5
   35  10  14
   7  4  2  5
```

Rather than computing the product of every possible path from the top to the bottom of the pyramid, we consider the prime factorization of $140 = 2\times2\times5\times7$.

The numbers in the pyramid are not all prime. So, we also consider the prime factorizations of the composite numbers in the pyramid: $35 = 5\times7$, $10 = 2\times5$, $14 = 2\times7$, and $4 = 2\times2$.

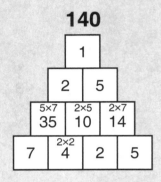

Now we look for a path of squares that includes exactly two 2's, one 5, and one 7, beginning with the 1 at the top of the pyramid.

We notice that no path that passes through the 10 block includes any 7's. So, our path will not pass through the 10 block.

Below are the four other possible paths. The only path with product $2\times2\times5\times7$ is shown by the numbers circled on the right.

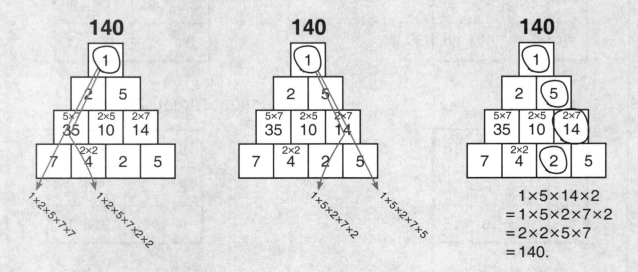

$1\times5\times14\times2$
$= 1\times5\times2\times7\times2$
$= 2\times2\times5\times7$
$= 140.$

26 Beast Academy Practice 4C

PRACTICE | Complete each Pyramid Descent puzzle below.

101.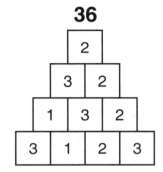

102.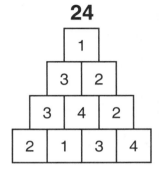

103.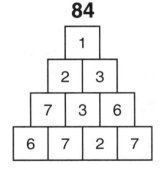

104.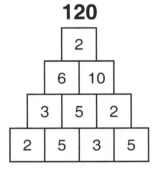

105.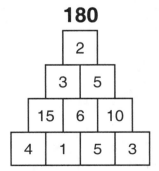

106.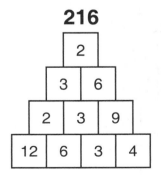

Pyramid Descent

PRACTICE | Complete each Pyramid Descent puzzle below.

107.
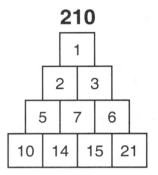

108.
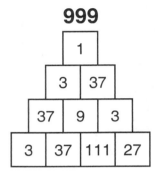

109.
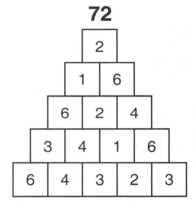

110.
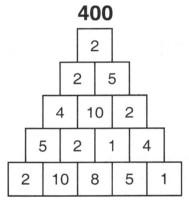

111. ★
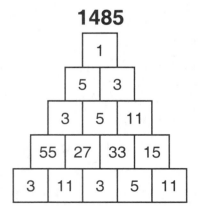

112. ★
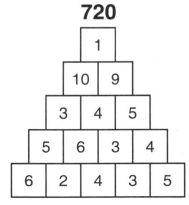

EXAMPLE | Is $17,640 = 2^3 \times 3^2 \times 5 \times 7^2$ divisible by 28?

Instead of dividing 17,640 by 28, we consider the prime factorizations of both numbers:

$17,640 = 2^3 \times 3^2 \times 5 \times 7^2$ and $28 = 2^2 \times 7$.

The prime factorization of 28 has two 2's and one 7. We look for two 2's and one 7 in the prime factorization of 17,640. This is easy to see when we write the prime factorization of 17,640 without exponents:

$17,640 = \boxed{2} \times \boxed{2} \times 2 \times 3 \times 3 \times 5 \times \boxed{7} \times 7$.

Rearranging and grouping $2 \times 2 \times 7$, we have

$17,640 = (2 \times 2 \times 7) \times 2 \times 3 \times 3 \times 5 \times 7$
$ = (28) \times 2 \times 3 \times 3 \times 5 \times 7$.

Therefore, 28 is a factor of 17,640.

Since 28 is a factor of 17,640, 17,640 **is divisible** by 28.

PRACTICE | Answer each question below.

113. Circle every number below that is a factor of $890,428 = 2^2 \times 7^3 \times 11 \times 59$.

    4   8   14   21   22   236

114. Circle every number below that has $1,323 = 3^3 \times 7^2$ as a factor.

$7,938 = 2 \times 3^4 \times 7^2$ $37,044 = 2^2 \times 3^3 \times 7^3$ $79,233 = 3 \times 7^4 \times 11$ $361,179 = 3^4 \times 7^3 \times 13$

115. List all of the factors of $480,200 = 2^3 \times 5^2 \times 7^4$ that are less than 20.

115. _____

116. ★ Given that $42,250 = 2 \times 5^3 \times 13^2$ and $325 = 5^2 \times 13$, what is $42,250 \div 325$?   **116.** _____

117. ★ ✎ Is $34,992 = 2^2 \times 3^5 \times 6^2$ divisible by $1,296 = 6^4$? Why or why not?

Divisibility with Prime Factorizations

Times Table Puzzles

These are just like the tables we used when learning multiplication facts!

The number in any white box is the product of the shaded numbers above and to the left of the box.

EXAMPLE In the times table on the right, some of the products have been filled in, but the factors are all missing. Fill in all of the missing entries.

×		
	42	70
	33	

We label the factors A, B, C, and D as shown.

The factor labeled A in the times table is a factor of both 42 and 70.
The factor labeled C in the times table is a factor of both 42 and 33.

×	C	D
A	42	70
B	33	

We write the prime factorizations of 42, 70, and 33:

$$42 = 2 \times 3 \times 7$$
$$70 = 2 \times 5 \times 7$$
$$33 = 3 \times 11$$

We know $A \times C = 42 = 2 \times 3 \times 7$.

×	C	D
A	2×3×7 42	2×5×7 70
B	3×11 33	

Since 2 and 7 are both factors of 42, but not of 33, we know 2×7 is a factor of A and not of C.

×	C	D
2×7 A	2×3×<u>7</u> 42	2×5×<u>7</u> 70
B	3×11 33	

Next, since 3 is a factor of 42, but not of 70, we know 3 is a factor of C and not of A.
Since 42 does not have any additional prime factors, we know that A and C do not have any additional prime factors.
So, $A = 2 \times 7 =$ **14** and $C =$ **3**.

×	3 **3**	D
2×7 **14**	2×<u>3</u>×7 42	2×5×7 70
B	3×11 33	

Then, $70 = (2 \times 7) \times D$. Since $70 = (2 \times 7) \times \boxed{5}$, we have $D =$ **5**.

Similarly, $33 = 3 \times B$. Since $33 = 3 \times \boxed{11}$, we have $B =$ **11**.

×	3 **3**	5 **5**
2×7 **14**	2×3×7 42	2×<u>5</u>×7 70
<u>11</u> **11**	3×<u>11</u> 33	

We multiply $B \times D = 11 \times 5 =$ **55** to get the missing product.

Our times table is complete, and we check to make sure all of the products are correct.

×	3 **3**	5 **5**
2×7 **14**	2×3×7 42	2×5×7 70
<u>11</u> **11**	3×11 33	5×11 **55**

PRACTICE | Solve each Times Table puzzle below.

118.

×		
	14	35
		15

119.

×		
	12	20
	33	

120.

×		
		84
	51	68

121.

×		
	65	91
		42

122.

×		
	165	88
	135	

123.

×		
		40
	63	105

Times Table Puzzles

PRACTICE | Solve each Times Table puzzle below.

124.

×		
	54	48
	63	

125.

×		
	144	200
	270	

126. ★

×			
	60	32	
		72	63

127. ★

×			
	88		68
		175	119

128. ★

×			
	56	98	
	36		90

129. ★

×			
	294	210	
	112		88

Factor Nim is a game for two players. Players take turns writing numbers. A game begins with the number 60. Each turn, a player takes the current number, subtracts one of its factors, and writes the result. The goal is to force your opponent to write the number 0.

For example, in the game below, Grogg plays first by subtracting 15 from 60 and writing 45.

$$\cancel{60} \quad 45$$

The second player then subtracts a factor of the new number and writes the result. In the game below, Alex subtracts 9 from 45 and writes 36.

$$\cancel{60} \quad \cancel{45} \quad 36$$

The game continues with players taking turns subtracting factors and writing the new result. In the completed game below, Grogg is forced to write 0. So, Alex is the winner.

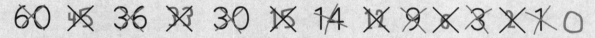

Variations:
- Players can choose to begin the game with any number.
- Start the game with **two** numbers. Each turn, a player chooses to subtract a factor from either number. As in the standard game, the first player who is forced to write the number 0 loses.

PRACTICE | After playing several games of Factor Nim, answer the questions below.

130. In a game of Factor Nim that begins with the number 4, is it better to play first or second? Explain your reasoning.

131. List all of the possible numbers you can get by subtracting a factor of 45 from 45. How many of these results are odd?

132. Is it possible to subtract a factor of an odd number to get an odd result? If so, name the odd number and its factor. If not, explain why it is impossible.

133. Find and describe a strategy that the first player can use to win a game of Factor Nim that begins with 60.

Beast Academy Practice 4C

Factors Challenge Problems

PRACTICE | Answer each question below.

134. Xavier adds the digits of a 1,000-digit number. The sum of the 1,000 digits is 4,878. Is Xavier's number divisible by 3, 9, both, or neither?

135. Your friend says, "I think 87,654,321 is prime!" Explain how you would convince your friend that 87,654,321 is not prime.

136. An exclamation point after a number tells us to multiply the number by each of the whole numbers less than it, all the way down to 1. For example, 5! = 5×4×3×2×1. Write the prime factorization of 7! using exponents.

136. _____

137. What is the smallest composite number that has four *different* prime factors?

137. _____

PRACTICE | Answer each question below.

138. There are two prime numbers between 4,020 and 4,030. **138.** _____ and _____
★ What are they?

139. Cammie says, "I think I have a divisibility test for 6. If a number is divisible
★ by 2 and by 3, then it is divisible by 6." Is she right? Why or why not?

140. Ralph says, "I think I have a divisibility test for 12. If a number is divisible
★ by 2 and by 6, then it is divisible by 12." Is he right? Why or why not?

141. What is the smallest composite number that is *not* divisible **141.** _____
★ by 2, 3, or 5?

Beast Academy Practice 4C

35

FACTORS
Challenge Problems

PRACTICE | Answer each question below.

142. Name a three-digit number that is divisible by 2, 3, 4, 5, 6, and 9. **142.** _____
★ *An extra challenge: There are 5 such numbers. Find them all!*

143. What is the smallest odd number that has three different **143.** _____
★ prime factors?

144. What is the smallest nonzero number that has both 16 and 28 **144.** _____
★ as factors?

Challenge Problems

PRACTICE | Answer each question below.

145. What is the largest number that is a factor of both 198 and 330?

145. _____

146. The 24 ways to arrange three different prime digits into a 3-digit number are shown below. All but two of these numbers are composite. For example, 327 = 3×109, 253 = 11×23, and 527 = 17×31. Which two of these 24 three-digit numbers are prime?

235	325	523	723
237	327	527	725
253	352	532	732
257	357	537	735
273	372	572	752
275	375	573	753

146. _____
_____

147. Lucas has three sisters: Cora, Dora, and Flora. Lucas tells Winnie, "The product of my sisters' ages is 36, and the sum of their ages is the same as my locker number. Can you figure out their ages?" Winnie looks at the number on Lucas's locker, then says, "I don't have enough information."
What is Lucas's locker number?

147. _____

Beast Academy Practice 4C

CHAPTER 8
Fractions

Use this Practice book with Guide 4C from BeastAcademy.com.

Recommended Sequence:

Book	Pages:
Guide:	50-57
Practice:	39-45
Guide:	58-67
Practice:	46-55
Guide:	68-77
Practice:	56-71

You may also read the entire chapter in the Guide before beginning the Practice chapter.

A *fraction* is a number.
Fractions are also another way to write division.
For example, we can write 5÷7 as $\frac{5}{7}$ or "five sevenths."

The 5 is the **numerator**. It's the number being divided.

The 7 is the **denominator**. It's the number we are dividing by.

We can locate $\frac{5}{7}$ on the number line by splitting the line between 0 and 1 into 7 equal pieces. Each piece has length $\frac{1}{7}$. We count five lengths of $\frac{1}{7}$ from 0 and mark $\frac{5}{7}$ at the end of the fifth piece.

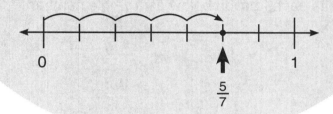

PRACTICE | Label the number marked on each number line below.

1.

2.

3.

4.

5.

6.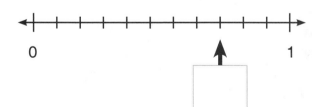

Beast Academy Practice 4C

FRACTIONS
Review

EXAMPLE | Write $\frac{65}{5}$ as a whole number.

Since fractions are another way to write division, $\frac{65}{5}$ means $65 \div 5$.

So, $\frac{65}{5}$ equals $65 \div 5 = $ **13**.

PRACTICE | Each fraction below is equal to a whole number. Write each fraction below as a whole number.

7. $\frac{48}{8} = $ _____

8. $\frac{91}{7} = $ _____

9. $\frac{55}{11} = $ _____

10. $\frac{75}{15} = $ _____

11. $\frac{156}{3} = $ _____

12. $\frac{52}{52} = $ _____

PRACTICE | For each equation below, fill in the numerator so that the equation is true.

13. $\frac{}{7} = 2$

14. $\frac{}{3} = 12$

15. $\frac{}{3} = 17$

16. $\frac{}{5} = 1$

17. $\frac{}{13} = 11$

18. $\frac{}{15} = 7$

FRACTIONS Review

EXAMPLE | Label $\frac{27}{4}$ on the number line below with a mixed number.

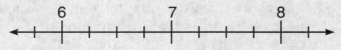

$\frac{27}{4}$ is between $\frac{24}{4} = 6$ and $\frac{28}{4} = 7$.

$\frac{27}{4}$ is three fourths more than $\frac{24}{4}$. So, written as a mixed number, $\frac{27}{4} = 6 + \frac{3}{4} = 6\frac{3}{4}$.

We count three fourths past 6 to reach $6\frac{3}{4} = \frac{27}{4}$.

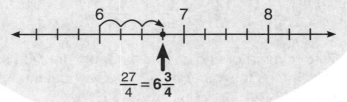

A mixed number is a whole number followed by a fraction that is less than 1.

PRACTICE

19. Label $\frac{16}{3}$ on the number line below as a mixed number.

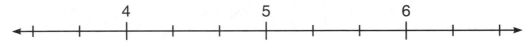

20. Label $\frac{55}{8}$ on the number line below as a mixed number.

21. Label $\frac{39}{4}$ on the number line below as a mixed number.

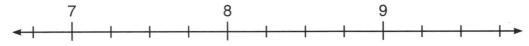

22. Label $\frac{34}{5}$ on the number line below as a mixed number.

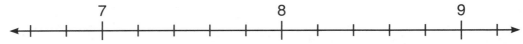

Fractions Review

> Fractions that represent the same number are equal and are called **equivalent fractions**.

EXAMPLE Write five fractions that are equivalent to $\frac{12}{30}$.

We can multiply both the numerator and denominator of a fraction by the same number to create an equivalent fraction. This is called **converting** the fraction.

$$\frac{12}{30} \xrightarrow{\times 2} \frac{24}{60} \qquad \frac{12}{30} \xrightarrow{\times 3} \frac{36}{90}$$

Similarly, we can divide both the numerator and denominator of a fraction by the same number to create an equivalent fraction. This is called **simplifying** the fraction.

$$\frac{12}{30} \xrightarrow{\div 2} \frac{6}{15} \qquad \frac{12}{30} \xrightarrow{\div 3} \frac{4}{10}$$

If the numerator and denominator of a fraction have no common factors other than 1, we say that the fraction is in **simplest form**.

$$\frac{12}{30} \xrightarrow{\div 2} \frac{6}{15} \xrightarrow{\div 3} \frac{2}{5}$$

So, $\frac{24}{60}$, $\frac{36}{90}$, $\frac{6}{15}$, $\frac{4}{10}$, and $\frac{2}{5}$ are all equivalent to $\frac{12}{30}$.

Of these, only $\frac{2}{5}$ is in simplest form.

> There are many other fractions equivalent to $\frac{12}{30}$!

PRACTICE For each equation below, fill in the blank numerator or denominator so that the equation is true.

23. $\dfrac{7}{9} = \dfrac{}{27}$

24. $\dfrac{12}{42} = \dfrac{2}{}$

25. $\dfrac{56}{40} = \dfrac{}{10}$

26. $\dfrac{6}{5} = \dfrac{72}{}$

27. $\dfrac{28}{42} = \dfrac{2}{} = \dfrac{}{9} = \dfrac{18}{}$

28. ★ $\dfrac{24}{30} = \dfrac{}{20}$

EXAMPLE | Compare $\frac{3}{8}$ to $\frac{15}{32}$.

To compare $\frac{3}{8}$ to $\frac{15}{32}$, we convert $\frac{3}{8}$ to an equivalent fraction with denominator 32. To do this, we multiply both the numerator and denominator of $\frac{3}{8}$ by 4.

$$\frac{3}{8} \overset{\times 4}{\underset{\times 4}{=}} \frac{12}{32}$$

$\frac{12}{32}$ is less than $\frac{15}{32}$. So, $\frac{3}{8}$ **is less than** $\frac{15}{32}$.

— or —

To compare $\frac{3}{8}$ to $\frac{15}{32}$, we convert $\frac{3}{8}$ to an equivalent fraction with numerator 15. To do this, we multiply both the numerator and denominator of $\frac{3}{8}$ by 5.

$$\frac{3}{8} \overset{\times 5}{\underset{\times 5}{=}} \frac{15}{40}$$

$\frac{1}{40}$ is less than $\frac{1}{32}$. So, $\frac{15}{40}$ is less than $\frac{15}{32}$. Therefore, $\frac{3}{8}$ **is less than** $\frac{15}{32}$.

We can also write $\frac{3}{8} < \frac{15}{32}$.

We can use equivalent fractions to help us make comparisons.

PRACTICE | Place <, >, or = in each circle to compare each pair of fractions below.

29. $\frac{3}{5} \bigcirc \frac{4}{5}$

30. $\frac{3}{8} \bigcirc \frac{3}{9}$

31. $\frac{3}{4} \bigcirc \frac{5}{8}$

32. $\frac{3}{5} \bigcirc \frac{11}{15}$

33. $\frac{12}{27} \bigcirc \frac{4}{9}$

34. $\frac{15}{23} \bigcirc \frac{5}{8}$

35. $\frac{40}{55} \bigcirc \frac{7}{11}$

36. $\frac{2}{3} \bigcirc \frac{32}{45}$

37. $\frac{45}{81} \bigcirc \frac{5}{7}$

38. $\frac{25}{39} \bigcirc \frac{8}{13}$

Beast Academy Practice 4C

FRACTIONS — Addition

EXAMPLE | What is $\frac{3}{7}+\frac{2}{7}$?

Adding 3 sevenths to 2 sevenths, we get a total of 3+2 = 5 sevenths.

So, $\frac{3}{7}+\frac{2}{7}=\frac{5}{7}$.

We can show this addition on the number line. Starting at $\frac{3}{7}$, we move right a distance of $\frac{2}{7}$ to $\frac{5}{7}$.

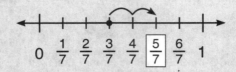

PRACTICE | Find the sum of each pair of fractions below. You may find it helpful to use the number line below each problem.

39. $\frac{4}{8}+\frac{1}{8}=$

40. $\frac{2}{5}+\frac{2}{5}=$

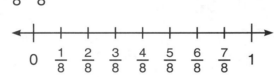

41. $\frac{2}{9}+\frac{5}{9}=$

42. $\frac{1}{6}+\frac{5}{6}=$

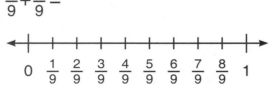

43. $\frac{4}{11}+\frac{3}{11}=$

44. $\frac{3}{10}+\frac{6}{10}=$

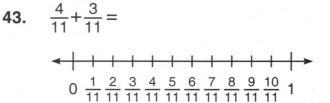

FRACTIONS: Addition

PRACTICE | Write each sum in simplest form.

45. $\frac{4}{6} + \frac{1}{6} =$

46. $\frac{1}{11} + \frac{1}{11} =$

47. $\frac{2}{5} + \frac{1}{5} =$

48. $\frac{4}{15} + \frac{8}{15} =$

49. $\frac{2}{7} + \frac{3}{7} =$

50. $\frac{7}{12} + \frac{1}{12} =$

51. $\frac{3}{8} + \frac{5}{8} =$

52. $\frac{2}{9} + \frac{4}{9} =$

PRACTICE | Answer each question below as a fraction in simplest form.

53. On the morning of April 5th, $\frac{1}{10}$ inches of rain fell on Beast Island. Later that day, another $\frac{7}{10}$ inches of rain fell. How many inches of rain fell on Beast Island on April 5th?

53. _____ in

54. Ally mixes $\frac{3}{8}$ cups of granola with $\frac{1}{8}$ cups of chocolate bits and $\frac{3}{8}$ cups of mixed nuts to make trail mix for her hiking trip. How many cups of trail mix does she make?

54. _____ c

FRACTIONS
Mixed Numbers

EXAMPLE Ms. Q. pours 64 ounces of lemonade into 5 glasses so that each glass holds an equal amount. How many ounces of lemonade are in each glass?

We can write the result of division as a fraction or a mixed number.

To find out how much lemonade is in each glass, we divide the total number of ounces by the number of glasses. $64 \div 5$ has quotient 12 and remainder 4.

After Ms. Q. pours 12 ounces in each glass, she has 4 ounces left to divide among the 5 glasses. So, each glass can receive an additional $4 \div 5 = \frac{4}{5}$ ounces of lemonade.

All together, there are $12 + \frac{4}{5} = \mathbf{12\frac{4}{5}}$ **ounces** of lemonade in each glass. We write:

$$64 \div 5 = \frac{64}{5} = \frac{60}{5} + \frac{4}{5} = 12 + \frac{4}{5} = \mathbf{12\frac{4}{5}}.$$

PRACTICE Answer each question below with a mixed number in simplest form.

55. A 15-ounce bag of potato chips contains 6 servings. How many ounces of chips are in each serving?

55. _____ oz

56. What is the side length in inches of a square with a perimeter of 39 inches?

56. _____ in

57. Grogg cuts a 22-inch rope into three pieces of equal length. How many inches long is each piece?

57. _____ in

58. Four identical bags of Beastie Puffs cereal weigh a total of 21 ounces. How many ounces does each bag of cereal weigh?

58. _____ oz

59. ★ Brenda makes some candles using 28 ounces of wax. She uses half the wax to make 2 large candles, and she uses the rest to make 4 identical small candles. How many ounces of wax are there in each small candle?

59. _____ oz

> We can convert any fraction greater than 1 into a mixed number.

EXAMPLE Write $\frac{57}{9}$ as a mixed number in simplest form.

$\frac{57}{9} = 57 \div 9$.

We consider dividing 57 ounces of clay among 9 little monsters. $57 \div 9$ has quotient 6 and remainder 3.

Each little monster gets 6 whole ounces of clay, with 3 ounces left to divide among the 9 monsters.

So, each monster gets another $3 \div 9 = \frac{3}{9}$ ounces of clay.

All together, each monster gets $6 + \frac{3}{9} = 6\frac{3}{9}$ ounces of clay.

We simplify $\frac{3}{9}$ to $\frac{1}{3}$, and we have $\frac{57}{9} = 6\frac{3}{9} = \mathbf{6\frac{1}{3}}$.

— or —

Since $6 = \frac{54}{9}$, we know $\frac{57}{9}$ is three ninths more than 6.

So, $\frac{57}{9} = \frac{54}{9} + \frac{3}{9} = 6 + \frac{3}{9} = 6\frac{3}{9} = \mathbf{6\frac{1}{3}}$.

Note that we could have begun by simplifying $\frac{57}{9}$ to $\frac{19}{3}$.

PRACTICE Write each fraction below as a mixed number in simplest form.

60. $\frac{53}{10} =$

61. $\frac{70}{9} =$

62. $\frac{35}{4} =$

63. $\frac{34}{8} =$

64. $\frac{92}{6} =$

65. $\frac{68}{12} =$

66. $\frac{137}{5} =$

67. $\frac{119}{14} =$

FRACTIONS
Mixed Numbers

PRACTICE | Answer each question below. You may find it useful to rewrite the fractions in each problem as mixed numbers.

68. How many whole numbers are between $\frac{39}{7}$ and $\frac{44}{3}$?   68. _____

69. Circle the fraction below that is closest to 10.

$\frac{79}{5}$ $\frac{17}{11}$ $\frac{111}{8}$ $\frac{51}{4}$

70. Write the four fractions below in order from least to greatest.   70. _____ _____ _____ _____

$\frac{59}{9}$ $\frac{27}{5}$ $\frac{45}{11}$ $\frac{31}{8}$

71. ★ Between which two consecutive whole numbers is $\frac{50}{6}+\frac{65}{7}$?   71. Between _____ and _____

72. ★ Place < or > in the circle to compare the expressions below.

$\frac{33}{4}+\frac{23}{7}$ ◯ $\frac{15}{2}+\frac{33}{16}$

Mixed Numbers

EXAMPLE | Write $4\frac{2}{5}$ as a fraction in simplest form.

$4\frac{2}{5}$ is two fifths more than 4, and $4 = \frac{20}{5}$.

So, $4\frac{2}{5} = 4 + \frac{2}{5} = \frac{20}{5} + \frac{2}{5} = \frac{22}{5}$.

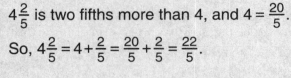

$\frac{22}{5}$ cannot be simplified.

Therefore, as a fraction in simplest form,

$$4\frac{2}{5} = \frac{22}{5}.$$

We can write any mixed number as a fraction!

PRACTICE | Write each mixed number as a fraction in simplest form.

73. $5\frac{1}{2} =$

74. $7\frac{2}{3} =$

75. $11\frac{6}{7} =$

76. $4\frac{4}{10} =$

77. $6\frac{4}{5} =$

78. $9\frac{4}{9} =$

Beast Academy Practice 4C — Guide Pages: 58-62 — 49

SKIP-COUNTING — FRACTIONS

EXAMPLE | Write the next four numbers in the skip-counting pattern below. Then, rewrite the pattern with the numbers in simplest form.

$$\frac{1}{10}, \frac{2}{10}, \frac{3}{10}, \square, \square, \square, \square$$

We add $\frac{1}{10}$ to each number to get the next number.

$$\frac{1}{10}, \frac{2}{10}, \frac{3}{10}, \boxed{\frac{4}{10}}, \boxed{\frac{5}{10}}, \boxed{\frac{6}{10}}, \boxed{\frac{7}{10}}$$

Then, we rewrite the pattern, simplifying when possible.

$$\frac{1}{10}, \frac{1}{5}, \frac{3}{10}, \frac{2}{5}, \frac{1}{2}, \frac{3}{5}, \frac{7}{10}$$

PRACTICE | Follow the directions to complete each skip-counting pattern below.

79. Count by elevenths starting at $\frac{1}{11}$.

$$\frac{1}{11}, \frac{2}{11}, \frac{3}{11}, \square, \square, \square, \square$$

80. Count by eighths starting at $\frac{1}{8}$.

$$\frac{1}{8}, \frac{2}{8}, \frac{3}{8}, \square, \square, \square, \square$$

On the line below, write all seven numbers in the sequence above in simplest form.

_____

81. Count by ninths starting at $\frac{4}{9}$.

$$\frac{4}{9}, \frac{5}{9}, \frac{6}{9}, \square, \square, \square, \square$$

On the line below, write all seven numbers in the sequence above in simplest form. Use whole or mixed numbers when possible.

_____

PRACTICE — Complete each skip-counting pattern below. Then, rewrite the pattern with the numbers in simplest form. Use whole or mixed numbers when possible.

82. Count by fourths starting at $\frac{3}{4}$.

$\frac{3}{4}$, $\frac{4}{4}$, $\frac{5}{4}$, ☐, ☐, ☐, ☐

On the line below, write the sequence above with each number in simplest form.

_____

83. Complete the skip-counting pattern below.

$\frac{3}{12}$, $\frac{5}{12}$, $\frac{7}{12}$, ☐, ☐, ☐, ☐

On the line below, write the sequence above with each number in simplest form.

_____

PRACTICE — Fill in the missing numbers in each skip-counting pattern below. Write each number in simplest form, using whole or mixed numbers when possible.

84. $\frac{2}{35}$, $\frac{3}{35}$, $\frac{4}{35}$, ☐, ☐, ☐, ☐

85. ★ $\frac{1}{6}$, $\frac{1}{3}$, ☐, ☐, $\frac{5}{6}$, ☐, ☐

86. ★★ $\frac{1}{3}$, $\frac{3}{5}$, $\frac{13}{15}$, ☐, ☐, ☐, ☐

Beast Academy Practice 4C

FRACTIONS — Maze Escape

In the mazes below, begin at the shaded number. Follow a skip-counting pattern to escape through the exit. You may only move up, down, left, or right to the next number.

PRACTICE | Find the path from the shaded square to the exit.

We begin at the shaded square marked $\frac{1}{8}$ and escape at the exit above the square marked $\frac{29}{8}$. We can move up, down, left, or right.

If we move down to $\frac{5}{8}$, we have made a jump of four eighths. We can continue adding $\frac{4}{8}$ to escape the maze:

$$\frac{1}{8}, \frac{5}{8}, \frac{9}{8}, \frac{13}{8}, \frac{17}{8}, \frac{21}{8}, \frac{25}{8}, \frac{29}{8}.$$

This is our escape path!
The complete path is shown to the right.

No other path is an escape route. For example, if we had instead moved left to $\frac{3}{8}$, we would try to continue adding two eighths to escape the maze:

$$\frac{1}{8}, \frac{3}{8}, \frac{5}{8}, \frac{7}{8}, \frac{9}{8}, \dots$$

No square above, below, or beside $\frac{7}{8}$ contains $\frac{9}{8}$, so this is not an escape route.

Find more of these puzzles at BeastAcademy.com!

PRACTICE | Begin at the shaded number and find the skip-counting path to the exit in each maze. There is only one escape path for each maze.

87.

$\frac{19}{20}$	$\frac{16}{20}$	$\frac{25}{20}$	$\frac{9}{20}$	$\frac{4}{20}$
$\frac{11}{20}$	$\frac{5}{20}$	$\frac{7}{20}$	$\frac{3}{20}$	$\frac{19}{20}$
$\frac{14}{20}$	$\frac{3}{20}$	$\frac{9}{20}$	$\frac{11}{20}$	$\frac{17}{20}$
$\frac{22}{20}$	$\frac{1}{20}$	$\frac{16}{20}$	$\frac{13}{20}$	$\frac{15}{20}$
$\frac{17}{20}$	$\frac{4}{20}$	$\frac{7}{20}$	$\frac{10}{20}$	$\frac{13}{20}$

88.

$\frac{44}{11}$	$\frac{30}{11}$	$\frac{16}{11}$	$\frac{20}{11}$	$\frac{30}{11}$
$\frac{26}{11}$	$\frac{21}{11}$	$\frac{15}{11}$	$\frac{9}{11}$	$\frac{40}{11}$
$\frac{33}{11}$	$\frac{27}{11}$	$\frac{35}{11}$	$\frac{18}{11}$	$\frac{50}{11}$
$\frac{39}{11}$	$\frac{45}{11}$	$\frac{63}{11}$	$\frac{69}{11}$	$\frac{60}{11}$
$\frac{22}{11}$	$\frac{51}{11}$	$\frac{57}{11}$	$\frac{75}{11}$	$\frac{70}{11}$

FRACTIONS
Maze Escape

Careful! Some of the fractions on this page have been simplified or converted to mixed numbers.

PRACTICE | Find the path from the shaded square to the exit.

89.

$\frac{20}{13}$	$\frac{22}{13}$	$\frac{24}{13}$	$\frac{26}{13}$	$\frac{28}{13}$
$\frac{18}{13}$	$\frac{36}{13}$	$\frac{42}{13}$	$\frac{32}{13}$	$\frac{30}{13}$
$\frac{24}{13}$	$\frac{30}{13}$	$\frac{48}{13}$	$\frac{90}{13}$	$\frac{96}{13}$
$\frac{28}{13}$	$\frac{60}{13}$	$\frac{54}{13}$	$\frac{84}{13}$	$\frac{102}{13}$
$\frac{30}{13}$	$\frac{66}{13}$	$\frac{72}{13}$	$\frac{78}{13}$	$\frac{80}{13}$

90.

$\frac{77}{19}$	$\frac{79}{19}$	$\frac{76}{19}$	$\frac{73}{19}$	$\frac{70}{19}$
$\frac{72}{19}$	$\frac{82}{19}$	$\frac{67}{19}$	$\frac{80}{19}$	$\frac{67}{19}$
$\frac{68}{19}$	$\frac{85}{19}$	$\frac{87}{19}$	$\frac{61}{19}$	$\frac{64}{19}$
$\frac{91}{19}$	$\frac{88}{19}$	$\frac{77}{19}$	$\frac{47}{19}$	$\frac{37}{19}$
$\frac{94}{19}$	$\frac{97}{19}$	$\frac{67}{19}$	$\frac{57}{19}$	$\frac{27}{19}$

91. ★

9	$7\frac{2}{3}$	$9\frac{1}{3}$	12	$12\frac{2}{3}$
$6\frac{2}{3}$	$7\frac{1}{3}$	8	$11\frac{1}{3}$	$13\frac{1}{3}$
$8\frac{2}{3}$	$8\frac{1}{3}$	$8\frac{2}{3}$	$10\frac{2}{3}$	14
$10\frac{2}{3}$	9	$9\frac{1}{3}$	10	$11\frac{2}{3}$
$12\frac{2}{3}$	$14\frac{2}{3}$	13	$8\frac{1}{3}$	12

92. ★

3	$3\frac{1}{7}$	$3\frac{2}{7}$	$3\frac{3}{7}$	$3\frac{4}{7}$
$2\frac{6}{7}$	$3\frac{6}{7}$	$4\frac{2}{7}$	$4\frac{5}{7}$	$3\frac{5}{7}$
$2\frac{5}{7}$	$3\frac{3}{7}$	$5\frac{2}{7}$	$5\frac{1}{7}$	$3\frac{6}{7}$
$2\frac{4}{7}$	3	$5\frac{5}{7}$	$5\frac{4}{7}$	6
$4\frac{3}{7}$	$7\frac{5}{7}$	$7\frac{2}{7}$	$6\frac{6}{7}$	$6\frac{3}{7}$

93. ★

$\frac{10}{3}$	$\frac{7}{3}$	$\frac{5}{3}$	$\frac{50}{9}$	$\frac{52}{9}$
$\frac{41}{9}$	$\frac{8}{3}$	$\frac{25}{9}$	$\frac{16}{3}$	5
$\frac{28}{9}$	$\frac{26}{9}$	$\frac{5}{3}$	$\frac{46}{9}$	$\frac{44}{9}$
$\frac{10}{3}$	$\frac{49}{9}$	$\frac{15}{3}$	$\frac{35}{9}$	$\frac{14}{3}$
$\frac{32}{9}$	$\frac{34}{9}$	4	$\frac{38}{9}$	$\frac{40}{9}$

94. ★

$\frac{28}{2}$	15	$\frac{63}{4}$	$\frac{33}{2}$	$\frac{31}{2}$
$\frac{25}{2}$	$\frac{57}{4}$	10	$\frac{69}{4}$	$\frac{21}{2}$
$\frac{55}{4}$	$\frac{27}{2}$	$\frac{39}{4}$	18	$\frac{81}{4}$
$\frac{47}{4}$	$\frac{51}{4}$	$\frac{21}{2}$	$\frac{75}{4}$	$\frac{39}{2}$
$\frac{26}{2}$	12	$\frac{45}{4}$	$\frac{83}{4}$	$\frac{35}{2}$

Beast Academy Practice 4C

Adding Mixed Numbers

EXAMPLE | Add $6\frac{5}{7}+2\frac{4}{7}$.

We could convert each mixed number to a fraction, add the fractions, then convert the result back to a mixed number.
Or, we could add the whole numbers and fractions separately:

$$6\frac{5}{7}+2\frac{4}{7} = \left(6+\frac{5}{7}\right)+\left(2+\frac{4}{7}\right)$$
$$= (6+2)+\left(\frac{5}{7}+\frac{4}{7}\right)$$
$$= 8+\frac{9}{7}.$$

Since the fractional part of $8+\frac{9}{7}=8\frac{9}{7}$ is not less than 1, $8\frac{9}{7}$ is not a valid mixed number. So, we cannot use $8\frac{9}{7}$ as our answer.

We write $\frac{9}{7}$ as $\frac{7}{7}+\frac{2}{7}=1+\frac{2}{7}$, and then continue adding.

$$6\frac{5}{7}+2\frac{4}{7} = \left(6+\frac{5}{7}\right)+\left(2+\frac{4}{7}\right)$$
$$= (6+2)+\left(\frac{5}{7}+\frac{4}{7}\right)$$
$$= 8+\frac{9}{7}$$
$$= 8+1+\frac{2}{7}$$
$$= 9+\frac{2}{7}$$
$$= 9\frac{2}{7}.$$

So, $6\frac{5}{7}+2\frac{4}{7}=\mathbf{9\frac{2}{7}}$.

If we get a fractional part that is greater than or equal to 1, we aren't done yet!

*The fractional part of a mixed number should always be **simplified and less than 1**.*

PRACTICE | Write each sum below as a whole or mixed number in simplest form.

95. $6\frac{5}{11}+4\frac{3}{11} =$

96. $5\frac{2}{7}+6\frac{3}{7} =$

97. $9\frac{4}{13}+3\frac{7}{13} =$

98. $2\frac{2}{5}+1\frac{1}{5} =$

99. $8\frac{6}{11}+4\frac{9}{11} =$

100. $2\frac{5}{9}+8\frac{4}{9} =$

Adding Mixed Numbers

PRACTICE | Write each sum below as a whole or mixed number in simplest form.

101. $3\frac{4}{5} + 2\frac{2}{5} =$

102. $3\frac{2}{5} + 9\frac{1}{5} =$

103. $8\frac{5}{6} + 3\frac{5}{6} =$

104. $4\frac{10}{19} + 25\frac{15}{19} =$

105. $17\frac{7}{8} + 2\frac{7}{8} + 2\frac{7}{8} =$

106. $6\frac{10}{13} + 22\frac{9}{13} + 8\frac{7}{13} =$

PRACTICE | Answer each question below as a mixed number in simplest form.

107. Mr. Jones uses $1\frac{2}{3}$ gallons of blue paint and $2\frac{2}{3}$ gallons of red paint to paint his garage. How many gallons of paint does Mr. Jones use all together?

107. _____ gal

108. What is the perimeter of an equilateral triangle whose sides are $2\frac{3}{4}$ inches long?

108. _____ in

109. What is the perimeter of a rectangle with width $5\frac{6}{7}$ inches and height $2\frac{4}{7}$ inches?

109. _____ in

Subtracting Fractions

EXAMPLE What is $\frac{7}{9} - \frac{5}{9}$?

Subtracting 5 ninths from 7 ninths leaves $7 - 5 = 2$ ninths.

So, $\frac{7}{9} - \frac{5}{9} = \frac{2}{9}$.

We can show this subtraction on the number line. Starting at $\frac{7}{9}$, we move left a distance of $\frac{5}{9}$ to $\frac{2}{9}$.

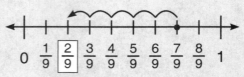

As with any subtraction problem, we can check our answer with addition:

$\boxed{\frac{2}{9}} + \frac{5}{9} = \frac{7}{9}$. ✓

We can show subtraction on the number line!

PRACTICE | Find each difference below. You may find it helpful to use the number line below each problem.

110. $\frac{4}{8} - \frac{1}{8} =$

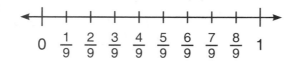

111. $\frac{3}{5} - \frac{2}{5} =$

112. $\frac{8}{9} - \frac{7}{9} =$

113. $\frac{5}{6} - \frac{4}{6} =$

114. $\frac{10}{11} - \frac{6}{11} =$

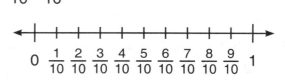

115. $\frac{8}{10} - \frac{5}{10} =$

56   Guide Pages: 68-72   Beast Academy Practice 4C

Subtracting Fractions

PRACTICE | Write each difference in simplest form.

116. $\dfrac{9}{13} - \dfrac{1}{13} =$

117. $\dfrac{8}{9} - \dfrac{1}{9} =$

118. $\dfrac{7}{12} - \dfrac{4}{12} =$

119. $\dfrac{7}{8} - \dfrac{5}{8} =$

120. $\dfrac{22}{15} - \dfrac{13}{15} =$

121. $\dfrac{20}{9} - \dfrac{14}{9} =$

PRACTICE | Answer each question below as a fraction in simplest form.

122. The weight of a dime on Beast Island is $\dfrac{2}{25}$ ounces. The weight of a Beast Island dollar coin is $\dfrac{7}{25}$ ounces. How many ounces heavier is a Beast Island dollar coin than a Beast Island dime?

    122. _____ oz

123. The distance from point A to point B on the line below is $\dfrac{25}{16}$ inches. The distance from point A to point C is $\dfrac{33}{16}$ inches. What is the distance in inches from point B to point C?

    123. _____ in

124. Alex swims $\dfrac{17}{10}$ miles in the morning. How many more miles must Alex swim in the afternoon if he wants to swim a total of $\dfrac{25}{10}$ miles for the day?

    124. _____ mi

Constellation Puzzles

In a **Constellation Puzzle**, the goal is to connect every group of equivalent expressions with a path of straight lines so that no two paths cross. Below is an example of a Constellation Puzzle and its solution.

Each set of equivalent expressions is connected by one continuous path using the following rules:

1. Dots on the paths must be connected by straight lines.
2. Paths may not cross.
3. When connecting two dots, you may not draw a line through a third dot.

When computing with fractions, we sometimes see numbers like $1\frac{4}{3}$ and $1\frac{5}{3}$.

*The fractional part of a mixed number should always be **less than 1**.*

A number like $1\frac{4}{3}$ or $1\frac{5}{3}$ should never be written as a final answer!

In this puzzle, we have $1\frac{4}{3} = \frac{7}{3} = 2\frac{1}{3}$ and $\frac{8}{3} = 1\frac{5}{3} = 2\frac{2}{3}$.

PRACTICE | Solve each Constellation Puzzle below.

125.

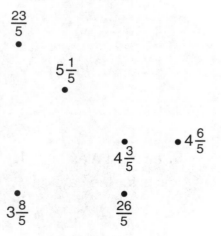

126.

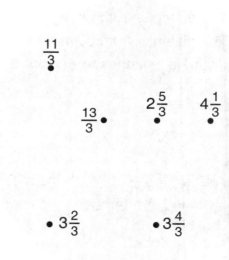

PRACTICE | Solve each Constellation Puzzle below.

127.

$1\frac{8}{5}$ $2\frac{4}{5}$

$\frac{16}{5}$ $2\frac{3}{5}$

$2\frac{6}{5}$

$\frac{14}{5}$

$\frac{13}{5}$ $3\frac{1}{5}$ $1\frac{9}{5}$

128.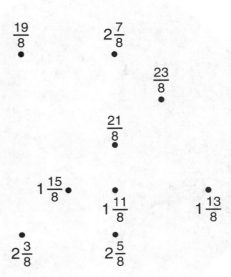

129.

$2\frac{14}{9}$ $3\frac{8}{9}$

$3\frac{2}{9}$

$3\frac{5}{9}$ $\frac{29}{9}$

$\frac{35}{9}$

$\frac{32}{9}$

$2\frac{11}{9}$ $2\frac{17}{9}$

130.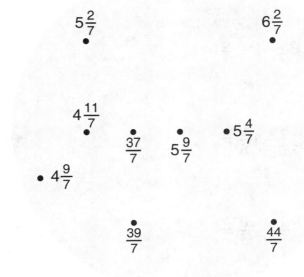

Subtracting Mixed Numbers

EXAMPLE | Subtract $7\frac{4}{5} - 3\frac{1}{5}$.

We could convert each mixed number to a fraction, then subtract:

$7\frac{4}{5} = \frac{35}{5} + \frac{4}{5} = \frac{39}{5}$ and $3\frac{1}{5} = \frac{15}{5} + \frac{1}{5} = \frac{16}{5}$.

So, $7\frac{4}{5} - 3\frac{1}{5} = \frac{39}{5} - \frac{16}{5} = \frac{23}{5}$.

Converting our final answer to a mixed number, we get $\frac{23}{5} = 4\frac{3}{5}$.

Therefore, $7\frac{4}{5} - 3\frac{1}{5} = \mathbf{4\frac{3}{5}}$.

— *or* —

We stack the mixed numbers and subtract without converting.

$\begin{array}{r} 7\frac{4}{5} \\ -3\frac{1}{5} \\ \hline 4\frac{3}{5} \end{array}$ $7 - 3 = 4$, and $\frac{4}{5} - \frac{1}{5} = \frac{3}{5}$.

So, $7\frac{4}{5} - 3\frac{1}{5} = \mathbf{4\frac{3}{5}}$.

We check our answer with addition:

$\boxed{4\frac{3}{5}} + 3\frac{1}{5} = 7\frac{4}{5}$. ✓

PRACTICE | Write each difference as a whole or mixed number in simplest form.

131. $8\frac{3}{7} - 2\frac{1}{7} =$

132. $9\frac{7}{11} - 5\frac{3}{11} =$

133. $3\frac{5}{9} - 1\frac{5}{9} =$

134. $7\frac{9}{10} - 2\frac{3}{10} =$

135. $5\frac{4}{5} - 2\frac{3}{5} =$

136. $6\frac{11}{15} - 4\frac{2}{15} =$

Subtracting Mixed Numbers

EXAMPLE | Subtract $6\frac{4}{7} - 2\frac{6}{7}$.

We could convert each mixed number to a fraction and subtract. Or, we stack the numbers and subtract without converting.

We cannot take away 6 sevenths from 4 sevenths.

$$\begin{array}{r} 6\frac{4}{7} \\ -2\frac{6}{7} \\ \hline \end{array}$$

So, we **regroup** $6\frac{4}{7}$. We take $1 = \frac{7}{7}$ from the 6 and add it to $\frac{4}{7}$:

$$6\frac{4}{7} = 5 + \frac{7}{7} + \frac{4}{7} = 5 + \frac{11}{7}.$$

So, $6\frac{4}{7} = 5\frac{11}{7}$. Now, we subtract.

$$\begin{array}{r} 6\frac{4}{7} \\ -2\frac{6}{7} \\ \hline \end{array} \longrightarrow \begin{array}{r} 5\frac{11}{7} \\ -2\frac{6}{7} \\ \hline 3\frac{5}{7} \end{array}$$

Therefore, $6\frac{4}{7} - 2\frac{6}{7} = \mathbf{3\frac{5}{7}}$.

We check our answer with addition:

$$\boxed{3\frac{5}{7}} + 2\frac{6}{7} = 6\frac{4}{7}. \checkmark$$

Sometimes subtracting the whole-number and fractional parts of numbers can get a little tricky.

PRACTICE | Write each difference as a mixed number in simplest form.

137. $10\frac{4}{9} - 6\frac{5}{9} =$

138. $4\frac{1}{3} - 2\frac{2}{3} =$

139. $7\frac{5}{11} - 3\frac{9}{11} =$

140. $8\frac{1}{4} - 3\frac{3}{4} =$

141. $8\frac{5}{14} - 6\frac{11}{14} =$

142. $12\frac{1}{9} - 5\frac{7}{9} =$

Mixed Number Arithmetic

FRACTIONS

EXAMPLE | Compute $10\frac{8}{9} + \frac{5}{9}$.

We could add the integer and fractional parts separately.

Or, to add five ninths to $10\frac{8}{9}$, we could start by adding one ninth to $10\frac{8}{9}$ to get a whole number. This gives us $10\frac{8}{9} + \frac{1}{9} = 11$.

Then, we add the remaining four ninths: $11 + \frac{4}{9} = 11\frac{4}{9}$.

We write $10\frac{8}{9} + \frac{5}{9} = 10\frac{8}{9} + \left(\frac{1}{9} + \frac{4}{9}\right)$
$= \left(10\frac{8}{9} + \frac{1}{9}\right) + \frac{4}{9}$
$= 11 + \frac{4}{9}$
$= \mathbf{11\frac{4}{9}}$.

PRACTICE | Express each sum as a mixed number in simplest form.

143. $2\frac{16}{17} + \frac{9}{17} =$

144. $4\frac{12}{13} + \frac{6}{13} =$

145. $9\frac{22}{23} + 8\frac{7}{23} =$

146. $7\frac{17}{19} + 3\frac{12}{19} =$

147. $5\frac{7}{12} + 2\frac{11}{12} =$

148. $9\frac{19}{21} + 3\frac{20}{21} =$

EXAMPLE | Compute $4\frac{1}{9} - 2\frac{8}{9}$.

FRACTIONS

Mixed Number Arithmetic

We count up from $2\frac{8}{9}$ to $4\frac{1}{9}$. From $2\frac{8}{9}$ to 3 is $\frac{1}{9}$. Then, from 3 to $4\frac{1}{9}$ is $1\frac{1}{9}$ more. So, the difference is $1\frac{1}{9} + \frac{1}{9} = \mathbf{1\frac{2}{9}}$.

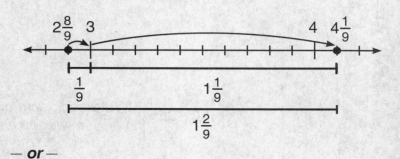

— *or* —

We can add any number to both numbers in a subtraction problem without changing their difference. We add $\frac{1}{9}$ to $4\frac{1}{9}$ and to $2\frac{8}{9}$ to make the subtraction easier.

$4\frac{1}{9} - 2\frac{8}{9}$ is equal to $\left(4\frac{1}{9} + \frac{1}{9}\right) - \left(2\frac{8}{9} + \frac{1}{9}\right) = 4\frac{2}{9} - 3 = \mathbf{1\frac{2}{9}}$.

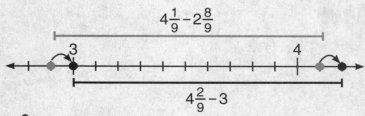

— *or* —

To subtract $2\frac{8}{9}$, we can take away 3 and give back $\frac{1}{9}$.

$$4\frac{1}{9} - 2\frac{8}{9} = 4\frac{1}{9} - 3 + \frac{1}{9}$$
$$= 1\frac{1}{9} + \frac{1}{9}$$
$$= \mathbf{1\frac{2}{9}}.$$

There are lots of ways to think about subtraction!

PRACTICE | Express each difference in simplest form, using mixed numbers when possible.

149. $4\frac{7}{15} - 2\frac{14}{15} =$

150. $9\frac{2}{11} - 5\frac{10}{11} =$

151. $6\frac{4}{7} - 3\frac{6}{7} =$

152. $4\frac{12}{17} - 3\frac{15}{17} =$

153. $5\frac{3}{8} - 1\frac{7}{8} =$

154. $12\frac{8}{13} - 2\frac{10}{13} =$

Beast Academy Practice 4C

PRACTICE | Write each sum or difference in simplest form. Write your answer as a mixed number when possible.

155. $5\frac{1}{3} + 16\frac{1}{3} =$

156. $7\frac{3}{10} - 2\frac{7}{10} =$

157. $67\frac{3}{8} + 14\frac{7}{8} =$

158. $\frac{271}{300} - \frac{201}{300} =$

159. $\frac{27}{50} + \frac{37}{50} =$

160. $32\frac{7}{11} + 84\frac{9}{11} =$

161. $17\frac{1}{18} - 5\frac{13}{18} =$

162. $74 - 33\frac{18}{19} =$

163. $526\frac{7}{8} - 213\frac{3}{8} =$

164. $421\frac{4}{9} + 82\frac{1}{9} =$

165. $12\frac{2}{9} + 12\frac{4}{9} - 5\frac{2}{9} =$

166. $20\frac{1}{3} + 10\frac{1}{3} - 10\frac{2}{3} - 5\frac{2}{3} =$

Find more practice problems at BeastAcademy.com!

PRACTICE | Answer each question below. Simplify all answers, and write your answer as a mixed number when possible. Remember to include units with your answer.

167. Anna has some juice in her mug. After she pours $3\frac{3}{4}$ more ounces of juice into her mug, she has a total of $10\frac{1}{4}$ ounces of juice in her mug. How many ounces of juice were originally in Anna's mug?

167. _____

168. ★ Fiona's hair is $31\frac{2}{3}$ inches longer than Grogg's, and Nellie's hair is $18\frac{1}{3}$ inches longer than Grogg's. Nellie's hair is 23 inches long. How many inches long is Fiona's hair?

168. _____

169. ★ Luke is $5\frac{1}{4}$ inches taller than Jess. The sum of Luke's and Jess's heights is $82\frac{1}{4}$ inches. How many inches tall is Jess?

169. _____

170. ★ Patrice made $19\frac{3}{5}$ ounces of butterscotch fudge for the Beast Academy Bake Sale, and Michael made $9\frac{4}{5}$ ounces more fudge than Patrice. Patrice and Michael combine their fudge and then divide it equally into 35 bags. How many ounces of fudge are in each bag?

170. _____

Fraction-Sum Link

In a **Fraction-Sum Link** puzzle, the goal is to connect pairs of numbers whose sum is a given target.

- Paths may only go up, down, left, or right through squares.
- Paths must begin and end at a number, but they may not pass *through* squares that contain numbers.
- Only one path may pass through each square.

Below is an example of a Fraction-Sum Link puzzle and its solution. Paths connect pairs of fractions whose sum is 2.

Target Sum: 2

$\frac{3}{7} + 1\frac{4}{7} = 2$

$1\frac{3}{7} + \frac{4}{7} = 2$

$\frac{2}{7} + 1\frac{5}{7} = 2$

$1\frac{1}{7} + \frac{6}{7} = 2$

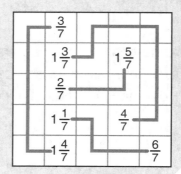

PRACTICE | Solve each Fraction-Sum Link puzzle below.

171. Target Sum: 3

	$1\frac{1}{5}$			
	$\frac{3}{5}$		$1\frac{3}{5}$	
$1\frac{4}{5}$	$\frac{1}{5}$			
		$1\frac{2}{5}$	$\frac{4}{5}$	
$2\frac{4}{5}$	$2\frac{1}{5}$			$2\frac{2}{5}$

172. Target Sum: 4

$1\frac{3}{7}$	$3\frac{3}{7}$	$2\frac{6}{7}$		
		$2\frac{4}{7}$	$3\frac{5}{7}$	
$\frac{2}{7}$	$\frac{4}{7}$			$1\frac{1}{7}$

PRACTICE | Solve each Fraction-Sum Link puzzle below.

173. ★ *Target Sum:* 5

	$2\frac{2}{5}$	$1\frac{4}{5}$	$2\frac{4}{5}$	$\frac{2}{5}$	
	$4\frac{3}{5}$		$2\frac{3}{5}$		
	$2\frac{1}{5}$				
$3\frac{1}{5}$					

174. ★ *Target Sum:* 6

				$4\frac{8}{9}$	
			$2\frac{4}{9}$	$5\frac{7}{9}$	
	$1\frac{4}{9}$		$4\frac{5}{9}$		
		$\frac{2}{9}$		$1\frac{1}{9}$	
					$3\frac{5}{9}$

175. ★ *Target Sum:* 5

$1\frac{4}{9}$						
		$1\frac{1}{9}$	$3\frac{5}{9}$	$2\frac{4}{9}$	$\frac{2}{9}$	
$3\frac{8}{9}$						
		$2\frac{5}{9}$				
					$1\frac{8}{9}$	$2\frac{7}{9}$
$4\frac{7}{9}$	$3\frac{1}{9}$	$\frac{5}{9}$			$4\frac{4}{9}$	
$2\frac{2}{9}$						

176. ★★ *Target Sum:* 7

		$3\frac{2}{3}$			
	$2\frac{2}{3}$			$1\frac{2}{3}$	
			$3\frac{1}{3}$	$6\frac{1}{3}$	
$4\frac{2}{3}$					
		$5\frac{1}{3}$			
	$\frac{2}{3}$			$2\frac{1}{3}$	
		$4\frac{1}{3}$			

Beast Academy Practice 4C   67

Longer Sums — FRACTIONS

EXAMPLE | Add $\frac{1}{7}+\frac{3}{7}+\frac{4}{7}+\frac{6}{7}$.

We can add all the numerators from left to right, put the result over 7, then simplify the sum.

Or, we can reorder and regroup the addition to make the sum easier to compute. In this expression, there are two pairs of fractions whose sum is 1:

$$\frac{1}{7}+\frac{3}{7}+\frac{4}{7}+\frac{6}{7}$$

So, we have:
$$\frac{1}{7}+\frac{3}{7}+\frac{4}{7}+\frac{6}{7} = \left(\frac{1}{7}+\frac{6}{7}\right)+\left(\frac{3}{7}+\frac{4}{7}\right)$$
$$= 1+1$$
$$= 2.$$

PRACTICE | Write each sum below as a whole number or a mixed number in simplest form.

177. $\frac{11}{23}+\frac{16}{23}+\frac{12}{23}=$

177. _____

178. $\frac{9}{11}+\frac{1}{11}+\frac{10}{11}=$

178. _____

179. $2\frac{2}{15}+\frac{8}{15}+1\frac{13}{15}=$

179. _____

180. $\frac{1}{7}+\frac{2}{7}+\frac{3}{7}+\frac{4}{7}+\frac{5}{7}+\frac{6}{7}=$

180. _____

FRACTIONS
Longer Sums

PRACTICE | Write the perimeter of each shape below as a whole or mixed number in simplest form. Remember to include units with your answer.

181. *Triangle*

Perimeter = _____

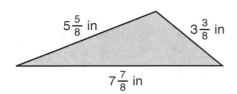

182. *Rectangle*

Perimeter = _____

183. *Pentagon*

Perimeter = _____

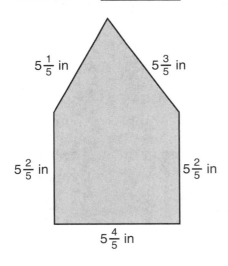

184. *Rectilinear hexagon*

Perimeter = _____

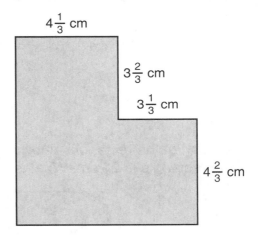

Beast Academy Practice 4C

FRACTIONS Challenge Problems

PRACTICE — Answer each question below. Simplify all answers, and write your answer as a mixed number when possible. Remember to include units when they are needed.

185. The height of a cube-shaped wood block is exactly $2\frac{3}{5}$ inches. What is the smallest nonzero number of these blocks that can be used to form a stack whose height is a whole number of inches?

185. _____

186. Compute

$$\left(3-\tfrac{1}{8}\right)+\left(3-\tfrac{2}{8}\right)+\left(3-\tfrac{3}{8}\right)+\left(3-\tfrac{4}{8}\right)+\left(3-\tfrac{5}{8}\right)+\left(3-\tfrac{6}{8}\right)+\left(3-\tfrac{7}{8}\right).$$

186. _____

187. ★ Yuan runs for ten days in a row. On the first day, she runs $\frac{5}{6}$ miles. Each day after the first, she runs $\frac{1}{6}$ miles farther than she did the day before. How many miles does Yuan run all together?

187. _____

188. ★ The width of the rectangle below is twice its height. The perimeter of the rectangle is 21 cm. What is its height in centimeters?

188. _____

Challenge Problems

PRACTICE — Answer each question below. Simplify all answers, and write your answer as a mixed number when possible. Remember to include units when they are needed.

189. ★ Two equilateral triangles are arranged as shown below to create a rhombus. The perimeter of each triangle is 8 inches. What is the perimeter of the rhombus in inches?

189. _____

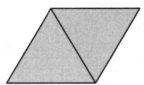

190. ★ Ralph is stacking cups. Each cup is $5\frac{3}{4}$ inches tall. Two stacked cups reach a height of $8\frac{1}{4}$ inches. How many inches tall is a stack of five cups?

190. _____

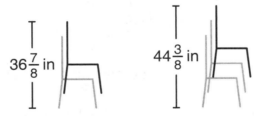

191. ★ A stack of two chairs is $36\frac{7}{8}$ inches tall. A stack of three chairs is $44\frac{3}{8}$ inches tall. How many inches tall is one chair on its own?

191. _____

192. ★ Joey's full glass of milk weighs 11 ounces. After Joey drinks half the milk, the glass and remaining milk together weigh $6\frac{3}{8}$ ounces. How many ounces does Joey's glass weigh on its own?

192. _____

Beast Academy Practice 4C 71

CHAPTER 9
Integers

Use this Practice book with
Guide 4C from BeastAcademy.com.

Recommended Sequence:

Book	Pages:
Guide:	80-93
Practice:	73-89
Guide:	94-109
Practice:	90-103

You may also read the entire chapter in the Guide before beginning the Practice chapter.

Negative numbers are numbers that are less than 0. They appear to the left of 0 on the number line.

-3

The negative symbol (-) tells us that the number is less than 0.

The 3 tells us how far it is to the left of 0.

So, -3 ("negative three") is the number **three units less than zero**.

INTEGERS
Negative Numbers

PRACTICE | Label the boxed missing numbers on each number line below.

1.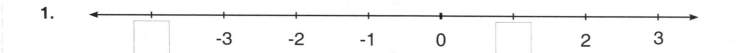

2.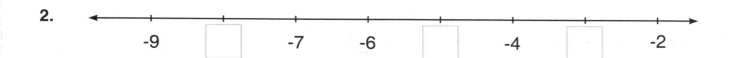

3.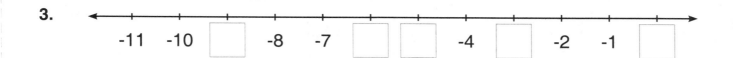

4.

5.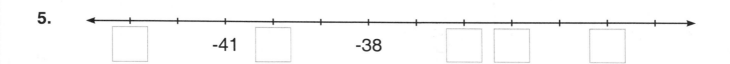

Beast Academy Practice 4C

INTEGERS: Temperatures

Negative numbers are often used to represent cold temperatures.

On a thermometer, negative temperatures appear below 0.

PRACTICE | Write the temperature indicated by each arrow on the thermometer below.

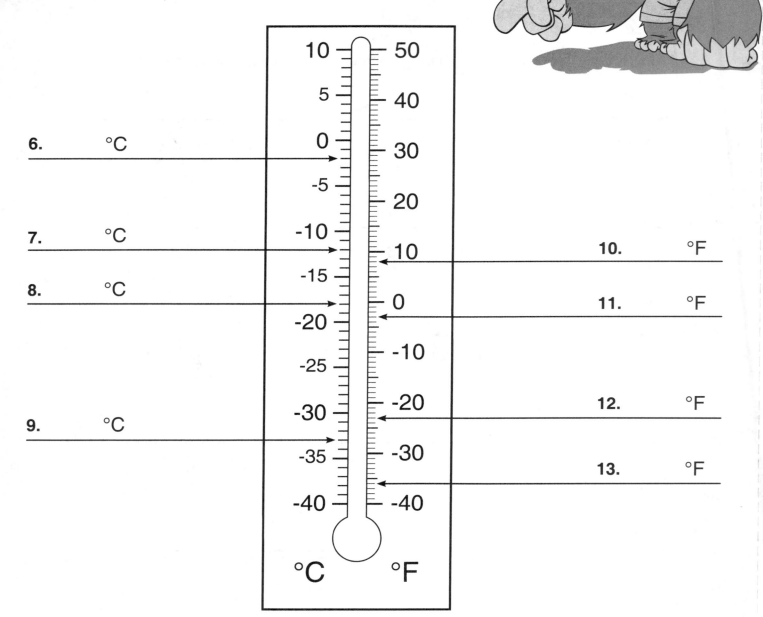

PRACTICE | Answer the following questions about temperatures. You may use the thermometer on the previous page.

14. What temperature is 15 degrees colder than 10°F?

14. _____ °F

15. One afternoon, the temperature at Yeti National Park was -1°C. Later in the evening, the temperature was -13°C. By how many degrees did the temperature drop?

15. _____

16. Write the four temperatures below in order from coldest to warmest.

    -9°F   -19°F   9°F   -29°F

16. _____, _____, _____, _____

17. Last night, the temperature at Beast Academy was -9°C. This morning, the temperature was 13 degrees warmer than it was last night. What was the morning temperature at Beast Academy?

17. _____ °C

18. The temperature at Lake Jackalope rose 22 degrees between midnight and 6 am. The temperature at 6 am was 10°C. What was the midnight temperature?

18. _____ °C

19. The temperature at the base of Mount Everbeast is -2°C. The temperature at the top is 35 degrees colder than at the base. What is the temperature at the top of Mount Everbeast?

19. _____ °C

Integers

Positive & Negative Integers

An *integer* is a number without a fractional part.

Positive integers are to the right of zero on the number line.

Negative integers are to the left of zero on the number line.

Nonnegative integers are all of the positive integers and zero. These are the integers that are **not negative**!

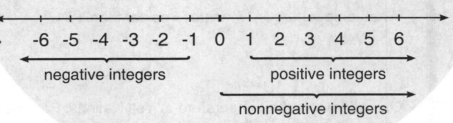

PRACTICE | Answer each question below.

```
<---|---|---|---|---|---|---|---|---|---|---|---|---|---|---|---|---|---|---|--->
   -9  -8  -7  -6  -5  -4  -3  -2  -1   0   1   2   3   4   5   6   7   8   9
```

20. How many positive integers are less than 7? 20. _____

21. How many negative integers are greater than -9? 21. _____

22. How many integers are less than 5 but greater than -5? 22. _____

23. Which two integers are 3 units from 0? 23. _____ and _____

24. Which two integers are 5 units from -2? 24. _____ and _____

25. How many units is -3 from 6? 25. _____

26. Which nonnegative integer is not a positive integer? 26. _____

EXAMPLE | Which is greater, -6 or -8?

Positive & Negative Integers

We look to the number line.

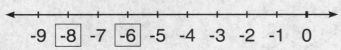

Since -6 is to the right of -8 on the number line, **-6 is greater than -8**.

With symbols, we can write
-8 ◁ -6 *or* -6 ▷ -8.

On the number line, the numbers to the left of -6 are less than -6.

Numbers to the right of -6 are greater than -6.

PRACTICE | Place a < or > in each circle below to compare each pair of numbers.

27. 3 ◯ -5

28. -4 ◯ 0

29. -5 ◯ -4

30. -3 ◯ -5

31. -25 ◯ -18

32. -412 ◯ -1,032

PRACTICE | Write each set of numbers below in order from least to greatest.

33. 3, -21, 0, -8

33. _____ < _____ < _____ < _____

34. -2, 12, -3, -9

34. _____ < _____ < _____ < _____

35. -11, -18, 8, -5

35. _____ < _____ < _____ < _____

36. -19, -2, -10, -7

36. _____ < _____ < _____ < _____

EXAMPLE | Trace a path in the hexagonal grid below that crosses all the numbers in the grid in order from least to greatest.

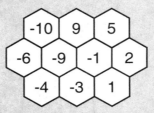

We begin by circling the smallest number on the grid, -10. Then, we move from hexagon to hexagon, always connecting to the next-smallest integer in the grid. We finish at the largest number on the grid, 9.

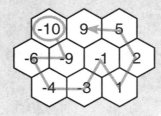

PRACTICE | Trace a path in each hexagonal grid below that crosses all the numbers in the grid in order from least to greatest.

37.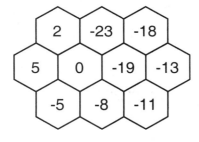

38.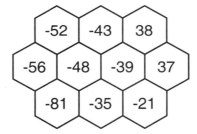

39.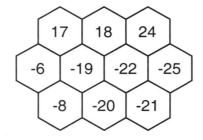

40.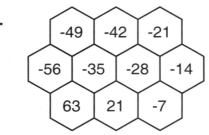

Integer Path Puzzles

PRACTICE | Trace a path in each hexagonal grid below that crosses all the numbers in the grid in order from least to greatest.

41.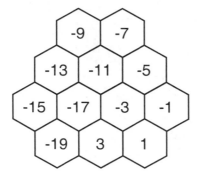

42.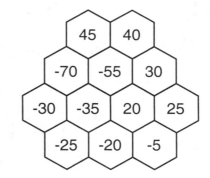

43.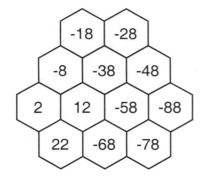

44.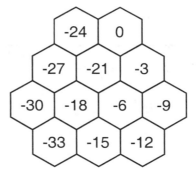

45.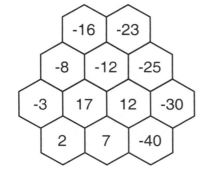

46.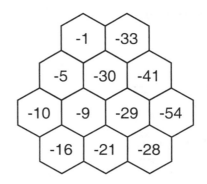

EXAMPLE | Compute -8+3.

To add 3 to any number on the number line, we start at the number and move 3 units to the right.

To add -8+3, we start at -8 and move 3 units to the *right*.

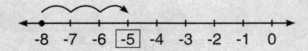

We arrive at -5. So, -8+3 = **-5**.

PRACTICE | Compute each sum below.

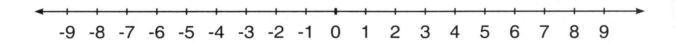

47. -9+4 = _____

48. -1+7 = _____

49. -7+7 = _____

50. -3+5 = _____

51. -8+13 = _____

52. -6+2 = _____

53. -8+7 = _____

54. -9+1 = _____

55. -1+5 = _____

56. -8+4 = _____

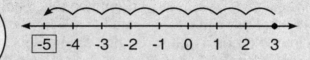

EXAMPLE | Compute 3+(-8).

To add -8 to any number on the number line, we start at the number and move 8 units to the **left**.

To add 3+(-8), we start at 3 and move 8 units to the left.

We arrive at -5. So, 3+(-8) = **-5**.

Note that this is the same result we get when adding -8+3, since addition is commutative.

When adding a positive number on the number line, we move right. But to add a negative, we do the opposite.

We move *left*.

PRACTICE | Compute each sum below.

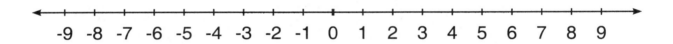

57. 6+(-4) = _____

58. -1+(-7) = _____

59. 2+(-6) = _____

60. -2+(-3) = _____

61. 9+(-3) = _____

62. -3+(-1) = _____

63. 8+(-11)+2 = _____

64. -6+3+(-4) = _____

65. 5+(-4)+(-1) = _____

66. -3+(-6)+3 = _____

Integer-Tac-Toe is a pencil-and-paper game for two players. Game play is similar to Tic-Tac-Toe, except that instead of placing X's and O's, players take turns placing integers.

On a standard 3-by-3 Tic-Tac-Toe board, the first player places a 1, 2, or 3 in any empty square. The second player then places a -1, -2, or -3 in an empty square. Play continues with the first player placing positives (1, 2, or 3) and the second player placing negatives (-1, -2, or -3) on the board.

The winner is the first player who creates a vertical (|), horizontal (–), or diagonal (╱ or ╲) line of three integers whose sum is 0. If every square is filled and no line of three integers has a sum of 0, the game ends in a draw.

In the sample game below, Winnie goes first and places positives. Grogg goes second and places negatives.

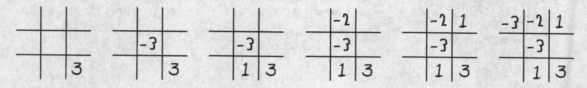

Winnie then places a 2 in the bottom-left square. The sum of the integers in the diagonal from top-right to bottom-left is 1+(-3)+2=0. So, Winnie wins!

Find a partner and play!

PRACTICE In the games below, Winnie is placing positives. Find Winnie's winning move by placing a **1, 2,** or **3** on each board to create a line of three integers whose sum is 0. Then, circle the winning line.

67.

	2	
-1	-1	
	3	

68.

1		3
-1		-2

69.

-2		
1	3	-2

70.

	-1	
-3	1	
3		

71.

	2	
	3	
	-2	-1

72.

3	-1	
1	-3	

Integer-Tac-Toe

PRACTICE — In the games below, Grogg is placing negatives. Find Grogg's winning move by placing a **-1, -2,** or **-3** on each board to create a line of three integers whose sum is 0. Then, circle the winning line.

73.
```
   |   | 2
---+---+---
   |-1 |
---+---+---
 2 | 1 |-1
```

74.
```
   |   | 3
---+---+---
   | 2 |-3
---+---+---
-1 |   | 2
```

75.
```
   | 1 |-1
---+---+---
   |   | 2
---+---+---
 3 |-1 |
```

76.
```
 1 |   |
---+---+---
 2 | 3 |-3
---+---+---
-1 |   |
```

77.
```
   |   |
---+---+---
 1 |-1 | 2
---+---+---
   |-3 | 2
```

78.
```
 2 |-1 | 1
---+---+---
   | 3 |
---+---+---
   |   |-1
```

PRACTICE — Answer each question below.

79. ★ ✏ The group (1, 1, -2) is a group of three integers that create a winning line in Integer-Tac-Toe, since -2+1+1=0. We consider (1, 1, -2) to be the same group as (1, -2, 1) and (-2, 1, 1). How many **different** groups of three integers create a winning line in Integer-Tac-Toe? List them below.

80. ★ It is Winnie's turn to play, and she must place a 1, 2, or 3 on the board to the right. Find the only move for Winnie that prevents Grogg from winning on his next turn.

```
 3 |-1 |
---+---+---
   | 1 |
---+---+---
   |-1 |
```

81. ★★ It is Grogg's turn to play, and he must place a -1, -2, or -3 on the board to the right. Find a play that will guarantee he can win after Winnie's next move.

```
 3 |-1 | 2
---+---+---
   |   |-1
---+---+---
   |   | 1
```

INTEGERS Addition Review

EXAMPLE | What is -38+25?

Consider the computation on the number line. We start at -38 and move 25 units to the right.

This brings us 25 units closer to zero. We end up 38−25 = 13 units to the left of zero at **-13**.

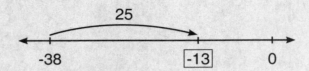

PRACTICE | Compute each sum below.

82. 19+(-11) = _____

83. (-11)+(-17) = _____

84. -20+(-13) = _____

85. -23+35 = _____

86. -12+(-71) = _____

87. 28+(-14) = _____

88. 25+16 = _____

89. 13+(-48) = _____

90. (-18)+18 = _____

91. 24+(-36) = _____

92. Which number is 35 more than -14? 92. _____

93. What do you get when you add 20 to the sum of -11 and -19? 93. _____

94. Without computing the sums below, circle those that are negative.

-16,987+6,654   -42,345+(-57,654)   856,915+(-532,812)   1,098,765+(-2,345,678)

You can find more practice problems at BeastAcademy.com!

EXAMPLE | Find the next three terms in the pattern below.

-9, -7, -5, -3, __, __, __

Each term in the pattern is 2 more than the one before it.

-9, -7, -5, -3, __, __, __ (+2 each)

So, we continue the pattern by adding 2's.

-9, -7, -5, -3, <u>-1</u>, <u>1</u>, <u>3</u>

PRACTICE | Fill in the blanks in each skip-counting pattern below.

95. -25, -20, -15, ____, ____, ____, ____, ____, ____

96. -53, -43, -33, ____, ____, ____, ____, ____, ____

97. -55, -42, -29, ____, ____, ____, ____, ____, ____

98. -16, ____, -12, -10, ____, ____, ____, ____, ____

99. ____, ____, -19, ____, ____, -10, ____, -4, ____

100. -24, ____, -14, ____, ____, ____, 6, ____, ____

101. -43, ____, ____, ____, -11, ____, ____, ____, 21

Sum Squares

In a **Sum Square** puzzle, the digits 1 through 9 are used to fill the nine squares in the grid, one digit per square.

Some of the numbers in the grid are positive, and some are negative.

The numbers above and to the left of the grid give the sum of the integers in each column and row. On the right is an example of a completed Sum Square.

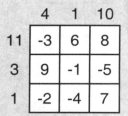

EXAMPLE | Complete the Sum Square puzzle below.

	5	12	0
6	1		
13			
-2	6	-5	-3

First, we look at the left column. We have 1+☐+6 = 5. This simplifies to 7+☐ = 5. Since 7+[-2] = 5, we fill the middle square of the left column with -2 as shown.

The remaining digits are 4, 7, 8, and 9.

	5	12	0
6	1		
13	-2		
-2	6	-5	-3

The two missing entries in the middle row must sum to 15 because -2+15 = 13. The only way to get a sum of 15 from two of the remaining digits is 7+8 = 15.

	5	12	0
6	1		
13	-2		
-2	6	-5	-3

Similarly, the two missing entries in the middle column must sum to 17 because -5+17 = 12. The only way to get a sum of 17 from two of the remaining digits is 8+9 = 17.

We learned above that the 8 is in the middle row. So, we place the 8 as shown in the center square, with the 9 above it.

	5	12	0
6	1	9	
13	-2	8	
-2	6	-5	-3

The remaining digits are 4 and 7.

In the top row, we have 1+9+[-4] = 6, and in the middle row, we have -2+8+[7] = 13.

All the digits have now been placed, and we check the sum of the integers in each row and column.

	5	12	0
6	1	9	-4
13	-2	8	7
-2	6	-5	-3

PRACTICE | Complete each Sum Square puzzle below.

102.

	16	9	20
12	1		
13	6	2	
20			7

103.

	3	-7	15
9	8	-6	
0		4	
2			9

104.

	9	0	0
9	6	5	
0			-7
0		-8	

105.

	-8	-10	-5
-4			8
-11		1	
-8	2		-4

106.

	0	0	3
0	1		-4
0		-9	
3	-8		

107.

	20	0	-11
12	9		-4
0		-2	
-3			-1

Beast Academy Practice 4C

INTEGERS
Sum Squares

PRACTICE | Complete each Sum Square puzzle below.

108.

	-1	2	8
1		-3	-5
3	-8		
5	-2		

109.

	-4	10	-11
4	-7		
-12		-4	
3		6	

110.

	7	7	7
1	9		-1
11			
9	-4		

111.

	8	-5	-2
-10	5		-7
10			
1		4	

112.

	-2	5	6
-1		2	-4
2			
8	-8		

113. ★

	0	-1	4
-4			
-9		-4	
16	9		-1

PRACTICE | Complete each Sum Square puzzle below.

114. ★

	-3	2	14
11	-5		
7		1	
-5			-3

115. ★

	6	7	-8
4	-5		
3			-9
-2		1	

116. What is the greatest possible sum of the numbers in a row or column of a Sum Square?

116. _____

117. What is the least possible sum of the numbers in a row or column of a Sum Square?

117. _____

118. ★★

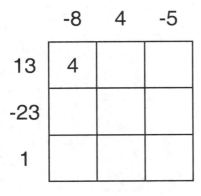

119. ★★

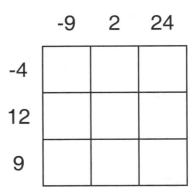

EXAMPLE | Compute 3−8.

To subtract 8 from any number on the number line, we start at the number and move 8 units to the **left**.

To subtract 3−8, we start at 3 and move 8 units to the left.

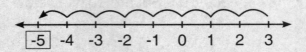

We arrive at -5. So, 3−8 = **-5**.

PRACTICE | Use the number line to help you compute each difference below.

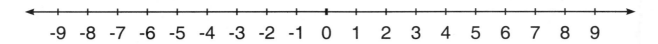

120. 2−7 = _____

121. -2−6 = _____

122. -4−3 = _____

123. 9−14 = _____

124. 5−11 = _____

125. 8−1 = _____

126. 6−10 = _____

127. -3−6 = _____

128. 0−4 = _____

129. -1−1 = _____

| EXAMPLE | Compute 3−(−8). |

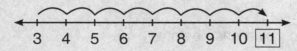

There are many ways to think about subtraction.

1. On the number line.

To subtract 8 from any number on the number line, we start at the number and move 8 units to the left. To subtract -8, we do the opposite. We move 8 units to the right. So, 3−(−8) = **11**.

2. Use addition to solve a subtraction problem.

We can treat any subtraction problem as a problem where we find the missing number in an addition equation.

For example, to compute 7−5 = ☐, we can find the number that fills the blank in ☐+5 = 7.

To compute 3−(−8) = ☐, we consider ☐+(−8) = 3. Since $\boxed{11}$+(−8) = 3, we know that 3−(−8) = **11**.

3. To subtract an integer, add its opposite.

Math beasts define subtraction in terms of addition. To subtract a number, we add its opposite. So, to subtract -8, we add 8.

So, we have 3−(−8) = 3+8 = **11**. This method is especially useful for subtracting a negative.

To subtract a number, we add its opposite.

| PRACTICE | Use Method 3 as described above to first write each subtraction problem below as an addition expression. Then, evaluate. |

Ex. −5−(−2) = __−5__ + __2__ = __−3__

130. 9−(−2) = ____ + ____ = ____

131. −6−(−3) = ____ + ____ = ____

132. 6−(−9) = ____ + ____ = ____

133. −5−11 = ____ + ____ = ____

134. −8−(−1) = ____ + ____ = ____

Subtracting Integers

> Sometimes, changing subtraction to addition makes the computation easier.

> Other times, it's easier to subtract without converting to addition.

PRACTICE | Compute each difference below.

135. 12 − 17 = _____

136. -11 − 15 = _____

137. 18 − (-6) = _____

138. -9 − (-14) = _____

139. -17 − (-7) = _____

140. 18 − 30 = _____

141. 0 − (-11) = _____

142. -13 − 13 = _____

143. -22 − (-14) = _____

144. -4 − (-4) = _____

145. -790 − 11 = _____

146. -23 − (-314) = _____

You can find more practice problems at BeastAcademy.com!

Subtracting Integers

EXAMPLE — Find two ways to arrange the integers 2, -3, and 5 in the blanks below to make a true equation.

☐ – ☐ = ☐.

There are six ways to arrange the three integers, as shown below:

2 – -3 = 5 ✓   2 – 5 = -3 ✓
-3 – 2 = 5 ✗   -3 – 5 = 2 ✗
5 – 2 = -3 ✗   5 – -3 = 2 ✗

Of these, only 2 – -3 = 5 and 2 – 5 = -3 are true statements.

PRACTICE — For each problem below, find two ways to arrange the given integers in the blanks to make a true equation.

147. **Integers:** -4, 5, 9   ☐ – ☐ = ☐ and ☐ – ☐ = ☐.

148. **Integers:** -5, 2, -3   ☐ – ☐ = ☐ and ☐ – ☐ = ☐.

149. **Integers:** -1, 3, 4   ☐ – ☐ = ☐ and ☐ – ☐ = ☐.

150. **Integers:** -5, 8, -13   ☐ – ☐ = ☐ and ☐ – ☐ = ☐.

151. **Integers:** -4, -14, -10 ☐ – ☐ = ☐ and ☐ – ☐ = ☐.

152. **Integers:** -7, -9, -16 ☐ – ☐ = ☐ and ☐ – ☐ = ☐.

INTEGERS Subtraction Patterns

EXAMPLE | Find the next three terms in the pattern below.

5, 2, -1, -4, ___, ___, ___

Each term in the pattern is 3 less than the one before it.

5, 2, -1, -4, ___, ___, ___ (−3, −3, −3)

So, we continue the pattern by subtracting 3's.

5, 2, -1, -4, **-7**, **-10**, **-13**

PRACTICE | Fill in the missing numbers in each pattern below.

153. 32, 22, 12, ____, ____, ____, ____, ____, ____

154. 17, 12, 7, ____, ____, ____, ____, ____, ____

155. 20, 9, -2, ____, ____, ____, ____, ____, ____

156. 11, ____, 3, -1, ____, ____, ____, ____, ____

157. ____, ____, -11, ____, ____, -35, ____, -51, ____

158. 29, ____, ____, -4, ____, ____, ____, -48, ____

159. ★ 13, ____, ____, ____, -15, ____, ____, ____, -43

Cross-Number Puzzles

EXAMPLE Fill in the missing entries in the Cross-Number puzzle below to make all the equations true.

	+	-2	=	-7
−		−		−
-6	+	5	=	
=		=		=
	+		=	

We fill in the missing entries as shown below.

$\boxed{-5}+(-2) = -7$. $-6+5 = \boxed{-1}$. $1+(-7) = \boxed{-6}$, **or**

$-2-5 = \boxed{-7}$. $-5-(-6) = \boxed{1}$. $-7-(-1) = \boxed{-6}$.

-5	+	-2	=	-7
−		−		−
-6	+	5	=	
=		=		=
	+	**-7**	=	

-5	+	-2	=	-7
−		−		−
-6	+	5	=	**-1**
=		=		=
1	+	-7	=	

-5	+	-2	=	-7
−		−		−
-6	+	5	=	-1
=		=		=
1	+	-7	=	**-6**

You can find more of these puzzles at BeastAcademy.com!

PRACTICE Fill in the missing entries in the Cross-Number puzzles below to make all the equations true.

160.

9	+	-2	=	
−		−		−
-3	+	5	=	
=		=		=
	+		=	

161.

-6	+		=	
−		−		−
2	+		=	15
=		=		=
	+	-9	=	

Beast Academy Practice 4C

95

Integers: Cross-Number Puzzles

PRACTICE — Fill in the missing entries in the Cross-Number puzzles below to make all the equations true.

162.

9	+		=	7
−		−		−
	+	-5	=	
=		=		=
	+		=	11

163.

-6	−	-2	=	
+		+		+
	−		=	
=		=		=
5	−	11	=	

164.

	+		=	4
−		−		−
5	+	-7	=	
=		=		=
	+	-2	=	

165.

	−	-5	=	
+		+		+
-5	−		=	2
=		=		=
7	−		=	

166.

	+	-6	=	2
−		−		−
	+		=	-3
=		=		=
7	+		=	

167.

	−	14	=	-21
+		+		+
-7	−		=	
=		=		=
	−		=	-35

Cross-Number Puzzles

PRACTICE — Fill in the missing entries in the Cross-Number puzzles below to make all the equations true.

168.

4	−		=	-2
+		−		+
-1	+	3	=	
=		=		=
	−		=	

169.

	+	-2	=	6
−		+		−
	−		=	-8
=		=		=
	+	3	=	

170.

-10	−	3	=	
+		−		+
	+		=	2
=		=		=
	−	-1	=	

171.

-3	+		=	
−		+		−
	−	5	=	
=		=		=
	+	1	=	-8

172.

5	+		=	-1
−		−		−
	+		=	
=		=		=
	+	-4	=	3

173.

	+		=	
−		+		−
	−	-2	=	5
=		=		=
-9	+		=	-4

Beast Academy Practice 4C — 97

INTEGERS — Absolute Value

The **absolute value** of a number is its distance from 0 on the number line.

$|-7|$

A pair of bars around a number is used to indicate absolute value.

Since -7 is 7 units from zero, the absolute value of -7 is 7.

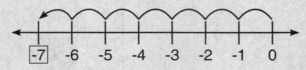

We can write this as an equation: $|-7| = 7$.

PRACTICE | Write the absolute value of each number.

174. $|26| =$ _____

175. $|-8| =$ _____

176. $|-35| =$ _____

177. $|72| =$ _____

PRACTICE | Use <, >, or = to compare each pair of expressions below.

178. $|-3|\ \bigcirc\ -4$

179. $|-17|\ \bigcirc\ |17|$

180. $|-54|\ \bigcirc\ |-19|$

181. $12\ \bigcirc\ |-13|$

PRACTICE | Answer each question below.

182. List every integer whose absolute value is 15.

182. _____

183. How many integers have absolute value 0?

183. _____

184. How many integers have absolute value -15?

184. _____

INTEGERS
Absolute Value

EXAMPLE | Compute $|-3+1|+2$.

First, we evaluate the part of the expression between the absolute value bars:

$$|-3+1|+2 = |-2|+2.$$

Then, we find the absolute value of -2. Finally, we add.

$$\begin{aligned}|-3+1|+2 &= |-2|+2 \\ &= 2+2 \\ &= \mathbf{4}.\end{aligned}$$

When absolute value bars appear around an expression, we first evaluate the expression. Then we find its absolute value.

PRACTICE | Evaluate each expression below.

185. $|-5-3| =$ _____

186. $|-5|-3 =$ _____

187. $|-10|-16 =$ _____

188. $|-10-16| =$ _____

189. $|2-5|-11 =$ _____

190. $2-|5-11| =$ _____

191. $|-13+3-14| =$ _____

192. $-13+|3-14| =$ _____

193. $|-6-3|-4 =$ _____

194. $-6-|3-4| =$ _____

Beast Academy Practice 4C

Absolute Value

EXAMPLE For which values of x is the equation below true?

$$|x+1|=2$$

The only two numbers with absolute value 2 are 2 and -2. So, $|x+1|=2$ tells us that $x+1$ is equal to 2 or -2.

If $x+1=2$, then $x=1$.

If $x+1=-2$, then $x=-3$.

So, there are two values of x for which $|x+1|=2$:

$x=$ **1** and $x=$ **-3**.

PRACTICE For each equation below, list all of the integers that could replace the x to make the equation true. If there is no integer that makes the equation true, write "impossible."

Ex. $|x|=5$ _x=5 and x=-5_

195. $4+|x|=5$ _____

196. $|x|=-4$ _____

197. $|x|-5=4$ _____

198. $|x|+5=4$ _____

199. $5-|x|=4$ _____

200. $4-|x|=5$ _____

201. $|x-4|=5$ _____

202. $|4-x|=5$ _____

Adding INTEGERS to Subtract

EXAMPLE | Compute 374+431−174.

Since subtracting a number is the same as adding its opposite, we can rewrite the problem above as an addition problem: 374+431+(-174).
This allows us to rearrange the terms, since addition is commutative and associative.

In our sum, there are two terms that end in 74: one positive and one negative. Since these two terms will be easiest to add, we rearrange to pair these terms: 374+(-174)+ 431.

Then, we evaluate:

$$\begin{aligned} 374+431-174 &= 374+431+(-174) \\ &= 374+(-174)+ 431 \\ &= \quad 200 \quad + 431 \\ &= \quad \textbf{631}. \end{aligned}$$

PRACTICE | Compute the value of each expression below.

203. 183+120−153 = _____

204. 159+217−49 = _____

205. 10+20+30−10−20 = _____

206. 7−6+8−7+9−8 = _____

207. 5+15+25+35−10−20−30 = _____

208. 28+48+68+88−18−28−38−48 = _____

209. -418+1,545+2,713−545−1,713+918 = _____

Beast Academy Practice 4C

INTEGERS Word Problems

PRACTICE | Answer each question below.

210. What is the sum of all the negative integers that are greater than -10? **210.** _____

211. What is the sum of all the integers that are greater than -4 but less than 4? **211.** _____

212. Which integer is the same distance from -13 as it is from 27 on the number line? **212.** _____

PRACTICE | Answer each question below.

213. Alex organizes the seven integers below into two groups so that the sum of the numbers in each group is the same.

    -9, -3, 11, 5, -4, -10, 14

    What is the sum of the numbers in one of Alex's two groups?

    213. _____

214. The sum of five consecutive integers is -5. What is the smallest of these five integers?

    214. _____

215. In a Magic Square, the sum of the numbers in each line of three numbers in a row, column, or diagonal is equal. In the Magic Square below, find an arrangement of the integers from -3 to 5 so that the sum of the numbers in each row, column, and diagonal equals 3.

    -3, -2, -1, 0, 1, 2, 3, 4, 5

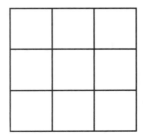

Beast Academy Practice 4C

103

HINTS
For Selected Problems

Below are hints to every problem marked with a ★.
Work on the problems for a while before looking at the hints.
The hint numbers match the problem numbers.

CHAPTER 7 — Factors

17. Since $59 \div n$ has a remainder of 3, what number must n divide with no remainder?
23. What types of numbers have an odd number of factors?
24. If a switch is flipped an even number of times, will it end up off or on? If a switch is flipped an odd number of times, will it end up off or on?
35. Is it possible to subtract two odd numbers and get 21?
57. Which set of three digits has a sum that is divisible by 9?
58. What is the smallest number of digits Grogg can use?
59. Which group(s) of three digits have a sum that is divisible by 3?
60. $\boxed{S\,T\,U}$, $\boxed{T\,U\,S}$, and $\boxed{U\,S\,T}$ are all even. What does that tell us about the digits S, T, and U?
63. Is it possible to arrange the digits 2, 3, 4, and 9 to make a 4-digit number that is *not* divisible by 3?
64. We must use at least one of each digit. How many 9's should we use? Then, how many 8's should we use?
97. Find the prime factorization of the target number and each number in the grid. Which number must be paired with the 75 in the bottom-left corner?
98. Find the prime factorization of the target number and each number in the grid. Which number must be paired with the 11 in the top-right corner?
99. Find the prime factorization of the target number and each number in the grid. Which number must be paired with the 16 in the bottom-right corner?
100. Find the prime factorization of the target number and each number in the grid. Which numbers must be grouped in a blob with the 2 in the bottom-right corner? There may be more than one way to draw a blob containing all these numbers.
111. $1,485 = 3^3 \times 5 \times 11$. There are seven squares that include at least one three (including $27 = 3 \times 3 \times 3$, $33 = 3 \times 11$, and $15 = 3 \times 5$). Find a path that passes through exactly three 3's.
112. $720 = 2^4 \times 3^2 \times 5$. There are four 5's in the pyramid (including $10 = 2 \times 5$). Which of these 5's must the path pass through?
116. We can solve a division problem using multiplication. What number can we multiply by $325 = 5^2 \times 13$ to get $42,250 = 2 \times 5^3 \times 13^2$?

117. 6 is not prime, so $2^2 \times 3^5 \times 6^2$ is not a prime factorization!
126. We can label the factors A, B, C, D, and E as shown.

×	C	D	E
A	60	32	
B		72	63

    3×3 is a factor of 72, but not of 32. What does this tell us about the prime factorization of B?

127. We can label the factors A, B, C, D, and E as shown.

×	C	D	E
A	88		68
B		175	119

    17 is a factor of 68, but not of 88. What does this tell us about the prime factorization of E?

128. We can label the factors A, B, C, D, and E as shown.

×	C	D	E
A	56	98	
B	36		90

    The prime factorization of $56 = 2 \times 2 \times 2 \times 7$ includes three 2's, but the prime factorization of $98 = 2 \times 7 \times 7$ includes only one 2. What does this tell us about the prime factorization of C?

129. We can label the factors A, B, C, D, and E as shown.

×	C	D	E
A	294	210	
B	112		88

    The prime factorization of $112 = 2^4 \times 7$ includes four 2's, but the prime factorization of $294 = 2 \times 3 \times 7^2$ includes only one 2. What does this tell us about the prime factorization of B?

130. Consider every possibility. What are the first player's options? Then, what are the second player's options?
131. Begin by listing the factors of 45. Without subtracting, can you tell whether each difference will be odd or even?
132. Can an odd number have an even factor?
133. Consider your previous answer. Why is this important?
138. Which whole numbers between 4,020 and 4,030 are *not* prime?
139. Consider the prime factorization of a number that is divisible by both 2 and 3.

140. Can you find a number that is divisible by both 2 and 6 that is not divisible by 12?

141. Consider the composite number's prime factorization. What is its smallest prime factor?

142. Consider the number's prime factorization. What is the smallest number of 2's, 3's, and 5's that must be included?

143. What primes are in the number's prime factorization?

144. Consider the prime factorizations of 16 and 28. What is the prime factorization of the smallest number that has both 16 and 28 as factors?

145. Consider the prime factorizations of 198 and 330. What prime factors do 198 and 330 have in common?

146. Which of these numbers are *not* prime?

147. Each sister's age must be a factor of 36. Organize a list of every possible set of three ages whose product is 36.

CHAPTER 8
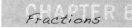
38

28. Can $\frac{24}{30}$ be simplified?

59. How many ounces of wax does Brenda use to make the small candles?

71. Write $\frac{50}{6}$ and $\frac{65}{7}$ as mixed numbers. Do you need to find an exact sum to know which whole numbers the sum is between?

72. Write each fraction as a mixed number. Which whole number you can compare both sums to?

85. Write all of the fractions with the same denominator.

86. Write all of the fractions with the same denominator.

91. The path begins by moving from $6\frac{2}{3}$ to $7\frac{1}{3}$.

92. The path begins by moving from $2\frac{4}{7}$ to 3.

93. Convert all of the numbers to fractions with the *same* denominator.

94. Convert all of the numbers to fractions with the *same* denominator.

168. How long is Grogg's hair?

169. Can you use the sum of Luke and Jess's heights to determine the height of the sum of Jess+Jess?

170. How much fudge did Michael make? How much fudge did Patrice and Michael make all together?

173. We have $2\frac{2}{5}+2\frac{3}{5}=5$. We connect this pair as shown.

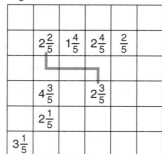

174. We have $4\frac{8}{9}+1\frac{1}{9}=6$. We connect this pair as shown.

175. We have $2\frac{2}{9}+2\frac{7}{9}=5$. We connect this pair as shown.

176. We have $4\frac{2}{3}+2\frac{1}{3}=7$. We connect this pair as shown.

187. Write a sum of 10 fractions to represent the total distance. How can you make this sum easier to evaluate?

188. If we let h represent the height of the rectangle, how could we represent the rectangle's width using h?

189. What is the length of one side of a triangle?

190. How many inches does each extra cup add to the height of the stack?

191. How many inches does each chair add to the height of the stack?

192. What is the weight of the milk that Joey drank?

CHAPTER 9
Integers
72

79. How can we organize our list to make sure we don't miss any groups of three integers?

80. First, determine where must Winnie place her number to block Grogg. Then, should she place a 1, a 2, or a 3?

81. Where can Grogg place his number to create two possible winning moves for his next turn?

113. Complete the bottom row, then the middle column. Then, which of the remaining digits cannot be placed in either the right column or the top row?

114. Consider all of the possibilities for the top row and for the right column. Which digit must be in both the top row and the right column?

115. Consider all of the possibilities for the middle column and for the top row. What digit must be placed in the top row, center column?

118. The sum of the numbers in the middle row is -23. The only way we can get this sum with three numbers in our puzzle is $(-6)+(-8)+(-9) = -23$. So, the middle row contains -6, -8, and -9, in some order.

Which of the remaining numbers must appear in the top row? The bottom row?

119. The sum of the numbers in the right column is 24. The only way we can get this sum with three numbers in our puzzle is $7+8+9 = 24$. So, the right column contains 7, 8, and 9, in some order.

Which of these numbers can be placed in the top right corner?

159. What number can we subtract four times to get from 13 to -15?

213. If Alex can group these numbers so that the two groups have an equal sum, how does the sum of each group relate to the sum of all seven integers?

214. Since we wish to find the smallest of the five integers, we use a variable to represent that integer: x. What expressions can we use to represent the next four consecutive integers?

— *or* —

We define m as the middle number of our five consecutive numbers. What expressions can we use to represent each of the four other integers?

215. Begin by looking for the center number.

We see that the number in the center of the Magic Square is part of four different sums.

A number in a corner is part of three different sums, while the remaining non-corner numbers are only part of two different sums.

If we search for all possible groups of three different integers from -3 to 5 whose sum is 3, only one of these numbers appears in at least four different sums.

— *or* —

Consider the sum of the numbers in the 9 shaded squares below:

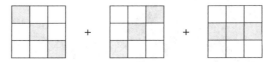

What is the sum of these numbers in the 9 shaded squares? Notice that the shaded squares include all six squares in the left and right columns, plus three copies of the center square, as shown below:

   (3 copies of the center number)

SOLUTIONS
Chapters 7-9

Chapter 7: Solutions

Introduction — page 7

1. Of these numbers, only 1, 2, 3, and 6 are factors of 18.
   (1) (2) (3) 4 5 (6) 7

2. Of these numbers, only 1, 3, and 7 are factors of 21.
   (1) (3) 5 6 (7) 14 42

3. Of these numbers, only 2, 9, 24, 36, and 72 are factors of 72.
   (2) 5 7 (9) (24) (36) (72)

4. Among these numbers, 5 is a factor of only 15, 30, and 45.
   12 (15) 21 (30) 38 (45) 53

5. Among these numbers, 3 is a factor of only 6, 36, and 45.
   2 (6) 16 22 (36) (45) 65

6. Among these numbers, 8 is a factor of only 16, 32, 48, and 96.
   4 12 (16) 22 (32) (48) (96)

7. The factors of 21 are 1, 3, 7, and 21. Of these numbers, only 1 and 3 are also factors of 45. So, the two numbers that are factors of both 45 and 21 are **1 and 3**.

   Notice that 1 is a factor of every number, since $n \div 1 = n$ for any number n.

Factor Pairs — 8-10

8. We begin by looking at the pairs with the smallest factors and organize our work by writing the smaller number in each pair first.

   30:
   1 × 30
   2 × 15
   3 × 10
   5 × 6

9. We list the factor pairs as shown below.

   56:
   1 × 56
   2 × 28
   4 × 14
   7 × 8

10. We list the factor pairs as shown below.

    35:
    1 × 35
    5 × 7

11. We list the factor pairs as shown below.

    54:
    1 × 54
    2 × 27
    3 × 18
    6 × 9

12. We list the factor pairs as shown below.

    26:
    1 × 26
    2 × 13

13. We list the factor pairs as shown below.

    40:
    1 × 40
    2 × 20
    4 × 10
    5 × 8

14. We first list the factor pairs as shown below.

    42:
    1 × 42
    2 × 21
    3 × 14
    6 × 7

    One factor in each of the four factor pairs is odd: 1, 3, 7, and 21. So, 42 has **4** odd factors.

15. We write out all the factors of 48 and 30 and compare.

48:	**30:**
1 × 48	1 × 30
2 × 24	2 × 15
3 × 16	3 × 10
4 × 12	5 × 6
6 × 8	

    The factors of 48 that are also factors of 30 are **1, 2, 3, and 6**.

16. To compute the rectangle's area, we multiply its width and height. Grogg's rectangle has whole-number side lengths. So, the number of inches in the width and height of his rectangle make a factor pair of 52.

    52:
    1 × 52
    2 × 26
    4 × 13

    52 has 6 factors, so Grogg's rectangle has one of **6** possible heights: 1, 2, 4, 13, 26, or 52 inches.

17. Since $59 \div n$ has a remainder of 3, we know that n must divide $59 - 3 = 56$ with no remainder. So, n is a factor of 56. We list these factors:

    56:
    1 × 56
    2 × 28
    4 × 14
    7 × 8

    $59 \div 1$ and $59 \div 2$ cannot have remainder 3, since the remainder must be less than the divisor. So, n cannot be 1 or 2.

    However, all other factors of 56 are greater than 3, so $59 \div 4$, $59 \div 7$, $59 \div 8$, $59 \div 14$, $59 \div 28$, and $59 \div 56$ all have remainder 3.

    So, n could be **4, 7, 8, 14, 28, or 56**.

18. We list the factor pairs as shown.

100:
1 × 100
2 × 50
4 × 25
5 × 20
10 × 10

19. We list the factor pairs as shown.

64:
1 × 64
2 × 32
4 × 16
8 × 8

20. The factors of 100 are 1, 2, 4, 5, 10, 20, 25, 50, and 100. So, 100 has **9** factors.

21. The factors of 64 are 1, 2, 4, 8, 16, 32, and 64. So, 64 has **7** factors.

22. Exactly one factor is repeated in a perfect square, as we saw in the previous problems. 100 = 10×10, so 10 is repeated in the factor pairs of 100, and 8×8 = 64, so 8 is repeated in the factor pairs of 64.

When using factor pairs to count the different numbers that are factors of a perfect square, we only count the repeated factor once. Every other pair contains two different numbers. All together, we count an odd number of different factors. **Therefore, every perfect square has an odd number of different factors.**

When we write the factor pairs of any number, we write an even number of factors. When the number is not a perfect square, all the factors in our pairs are *different*. So, numbers that are not perfect squares have an even number of factors. **Only perfect squares have an odd number of factors.**

23. As we found in the previous problem, only perfect squares have an odd number of factors.

Therefore, to find the numbers less than 50 that have exactly three (an odd number) factors, we look at the perfect squares between 4 and 50.

9:   **16:**   **25:**   **36:**   **49:**
1 × 9   1 × 16   1 × 25   1 × 36   1 × 49
3 × 3   2 × 8   5 × 5   2 × 18   7 × 7
             4 × 4   3 × 12
                                          4 × 9
                                          6 × 6

9, 25, and 49 are the three numbers between 4 and 50 that have exactly three factors.

24. The switches all start "off." A switch that is flipped an even number of times will be turned "off" as many times as it is turned "on," so it will end up in the "off" position. A switch that is flipped an odd number of times will end in the "on" position.

We know that every friend flips the switches whose numbers are multiples of his or her number. In other words, a switch n will be flipped by a friend if his or her number is a factor of n. For example, the factors of 15 are 1, 3, 5, and 15. So, switch 15 will be flipped by friends 1, 3, 5, and 15.

Therefore, to figure out whether a switch ends up off or on, we only need to know about the switch number's factors. In particular, we only need to know whether it has an *odd or even* number of factors.

For example, 15 has four factors: 1, 3, 5, and 15. Switch 15 is flipped on by friend 1, off by friend 3, on by friend 5, and off by friend 15.

However, 81 has five factors: 1, 3, 9, 27, and 81. Switch 81 is flipped on by friend 1, off by friend 3, on by friend 9, off by friend 27, and on by friend 81.

Only the switches numbered with perfect squares have an odd number of factors. So, only the switches numbered with perfect squares will be flipped an odd number of times and will end up "on."

So, the switches that end up in the "on" position are **1, 4, 9, 16, 25, 36, 49, 64, 81, and 100**.

25. We list the factor pairs of 39.

39:
1 × 39
3 × 13

39 has four factors, so 39 is **composite**.

26. We list the factor pairs of 47.

47:
1 × 47

47 has exactly two factors, so 47 is **prime**.

27. Zero is even, but is neither prime nor composite. The smallest even number greater than zero is 2. 2 has only two factors: 1 and 2. Therefore, 2 is prime.

Every even number has 2 as a factor, so each even number greater than 2 has at least three factors: 1, 2, and itself. So, no other even number is prime. Therefore, there is only **1** even prime.

28. We notice that 40 and 1,728 are both even. Every even number is divisible by 2. So, 40 and 1,728 each have at least three factors (1, 2, and itself). Therefore, 40 and 1,728 are not prime.

Then, we notice that 65 ends in a 5. Every number that ends in 0 or 5 is divisible by 5. So, 65 must be divisible by 5, and 65 has at least three factors (1, 5, and 65). Therefore, 65 is not prime.

We look at the factor pairs for the remaining numbers:

41:   **57:**   **63:**   **79:**
1 × 41   1 × 57   1 × 63   1 × 79
                 3 × 19   3 × 21
                                  7 × 9

Only 41 and 79 have exactly two factors, so they are the only two numbers in this group that are prime.

40   (41)   57   63   65   (79)   1,728

29. We notice that 42 is even. Every even number is divisible by 2. So, 42 has at least three factors (1, 2, and 42). Therefore, 42 is composite.

Then, we notice that 395 ends in a 5. Every number that ends in a 0 or 5 is divisible by 5. So, 395 has at least three factors (1, 5, and 395). Therefore, 395 is composite.

We look at the factor pairs for the remaining numbers:

29:	**37:**	**49:**	**51:**	**59:**
1×29	1×37	1×49	1×51	1×59
		7×7	3×17	

Only 42, 49, 51, and 395 have more than two factors, so they are the only four numbers in this group that are composite.

29   37   (42)   (49)   (51)   59   (395)

30. 1 is neither prime nor composite.

    2 has only two factors: 1 and 2. So, 2 is prime.

    As we saw in a previous problem, no other even number is prime. So, the remaining primes are odd.

    3, 5, and 7 each have just two factors, so they are prime.
    9 = 3×3, so 9 is composite.
    11 and 13 each have just two factors, so they are prime.
    15 = 3×5, so 15 is composite.
    17 and 19 each have just two factors, so they are prime.
    21 = 3×7, so 21 is composite.
    23 has exactly two factors, so it is prime.

    So, the primes less than 25 are **2, 3, 5, 7, 11, 13, 17, 19, and 23**.

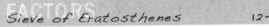

31. We show the grid each time we complete Steps 1 and 2:

    After circling 11, we notice that there are no multiples of 11 left to cross out. This is because every multiple of 11 that is less than 11×11 = 121 was crossed out when we crossed out multiples of 2, 3, 5, and 7. Similarly, there are no multiples of 13, 17, or any other uncircled number in the grid left to cross out.

    Now, every number in the grid is either circled or crossed out. Our final grid looks like this:

    The circled numbers are the primes less than 100.

32. The numbers we circled in the grid in the previous problem are 2, 3, 5, 7, 11, 13, 17, 19, 23, 29, 31, 37, 41, 43, 47, 53, 59, 61, 67, 71, 73, 79, 83, 89, and 97. We count **25** circled primes that are less than 100.

33. The largest two-digit prime is **97**.

34. Among the 21 two-digit primes, five have units digit 1, six have units digit 3, five have units digit 7, and five have units digit 9. So, **3** is the most common units digit among two-digit primes.

35. The difference between Lizzie's two primes is 21. The difference between any two odd numbers or any two even numbers is even. For example, 31−17 = 14, and 36−14 = 22. The difference between an odd and an even number is odd. For example, 67−12 = 55.

    Since Lizzie's difference is odd, she must have picked one odd prime and one even prime.

    We know that 2 is the only even prime. So, one of Lizzie's primes is [2].

    Then, [23]−2 = 21, so Lizzie's other prime is 23.

    When Lizzie adds the same two primes, her sum is 23+2 = **25**.

36. We look at our completed grid. The pairs of two-digit twin primes are (11 & 13), (17 & 19), (29 & 31), (41 & 43), (59 & 61), and (71 & 73).

    We count **6** pairs of two-digit twin primes.

37. We notice that when the digits of any prime with an even tens digit are reversed, the result is an even two-digit number. Since 2 is the only even prime, no even two-digit numbers are prime! So, we only look at two-digit primes that have *odd* tens digits.

    We list those primes and the result when their digits are reversed. We box the pairs whose digit reversals are a *different* prime number:

11 & 11	53 & 35
13 & 31	59 & 95
17 & 71	71 & 17
19 & 91	73 & 37
31 & 13	79 & 97
37 & 73	97 & 79

    Each pair appears twice in the boxes above. So, we have the following four pairs of two-digit emirps: **(13 & 31), (17 & 71), (37 & 73), and (79 & 97)**.

38. To find the Sophie Germain primes that are less than 50, we look at the result when we multiply each prime that is less than 50 by 2 and then add 1. We box the multiplied prime when the result is also prime.

$(2\times\boxed{2})+1=4+1=5.$ $(2\times\boxed{23})+1=46+1=47.$
$(2\times\boxed{3})+1=6+1=7.$ $(2\times\boxed{29})+1=58+1=59.$
$(2\times\boxed{5})+1=10+1=11.$ $(2\times 31)+1=62+1=63.$
$(2\times 7)+1=14+1=15.$ $(2\times 37)+1=74+1=75.$
$(2\times\boxed{11})+1=22+1=23.$ $(2\times\boxed{41})+1=82+1=83.$
$(2\times 13)+1=26+1=27.$ $(2\times 43)+1=86+1=87.$
$(2\times 17)+1=34+1=35.$ $(2\times 47)+1=94+1=95.$
$(2\times 19)+1=38+1=39.$

So, the Sophie Germain primes less than 50 are **2, 3, 5, 11, 23, 29, and 41**.

39. We subtract 1 from the first few multiples of 3. We box our result when it is prime.

$3-1=\boxed{2}.$ $6-1=\boxed{5}.$ $9-1=8.$

We notice the result of subtracting 1 from an odd number is always even. 2 is the only even prime. So, we will not find any other Eisenstein primes by subtracting 1 from an odd multiple of 3.

We consider only even multiples of 3 from now on:

$12-1=\boxed{11}.$ $42-1=\boxed{41}.$ $72-1=\boxed{71}.$
$18-1=\boxed{17}.$ $48-1=\boxed{47}.$ $78-1=77.$
$24-1=\boxed{23}.$ $54-1=\boxed{53}.$ $84-1=\boxed{83}.$
$30-1=\boxed{29}.$ $60-1=\boxed{59}.$ $90-1=\boxed{89}.$
$36-1=35.$ $66-1=65.$ $96-1=95.$

The next-largest even multiple of 3 is 102, but 102−1 is not less than 100.

So, the Eisenstein primes less than 100 are **2, 5, 11, 17, 23, 29, 41, 47, 53, 59, 71, 83, and 89**.

40. We subtract 1 from the first few powers of 2. We box our result when it is prime.

$2^1-1=2-1=1.$ $2^4-1=16-1=15.$
$2^2-1=4-1=\boxed{3}.$ $2^5-1=32-1=\boxed{31}.$
$2^3-1=8-1=\boxed{7}.$ $2^6-1=64-1=63.$

The next-largest power of 2 is $2^7=128$, but 128−1 is not less than 100. So, the Mersenne primes less than 100 are **3, 7, and 31**.

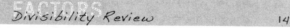

41. We look at the five numbers separately.

We begin with 7,007.
$7,007 = 7,000+7$
$= (7\times 1,000)+(7\times 1)$
$= 7\times(1,000+1).$
So, 7,007 is divisible by 7.

Then, we look at 49,049.
$49,049 = 49,000+49$
$= (7\times 7,000)+(7\times 7)$
$= 7\times(7,000+7).$
So, 49,049 is divisible by 7.

Next, we look at 147,285.
$147,285 = 140,000+7,000+280+5$
$= (7\times 20,000)+(7\times 1,000)+(7\times 40)+5$
$= 7\times(20,000+1,000+40)+5.$
147,285 is 5 more than a multiple of 7.
Therefore, 147,285 is not divisible by 7.

Then, we look at 700,721.
$700,721 = 700,000+700+21$
$= (7\times 100,000)+(7\times 100)+(7\times 3)$
$= 7\times(100,000+100+3).$
So, 700,721 is divisible by 7.

Finally, we look at 35,035,353.
$35,035,353 = 35,000,000+35,000+350+3.$
$= (7\times 5,000,000)+(7\times 5,000)+(7\times 50)+3$
$= 7\times(5,000,000+5,000+50)+3.$
35,035,353 is 3 more than a multiple of 7.
Therefore, 35,035,353 is not divisible by 7.

We circle the numbers that are divisible by 7.

⟨7,007⟩ ⟨49,049⟩   147,285   ⟨700,721⟩   35,035,353

42. We look at the five numbers separately.

We begin with 4,828.
$4,828 = 4,800+28$
$= 4,800+24+4$
$= (8\times 600)+(8\times 3)+4$
$= 8\times(600+3)+4.$
4,828 is 4 more than a multiple of 8.
Therefore, 4,828 is not divisible by 8.

Then, we look at 8,064.
$8,064 = 8,000+64$
$= (8\times 1,000)+(8\times 8)$
$= 8\times(1,000+8).$
So, 8,064 is divisible by 8.

Next, we look at 80,402.
$80,402 = 80,000+400+2$
$= (8\times 10,000)+(8\times 50)+2$
$= 8\times(10,000+50)+2.$
80,402 is 2 more than a multiple of 8.
Therefore 80,402 is not divisible by 8.

Then, we look at 484,560.
$484,560 = 480,000+4,000+560$
$= (8\times 60,000)+(8\times 500)+(8\times 70)$
$= 8\times(60,000+500+70).$
So, 484,560 is divisible by 8.

Finally, we look at 484,024.
$484,024 = 480,000+4,000+24$
$= (8\times 60,000)+(8\times 500)+(8\times 3)$
$= 8\times(60,000+500+3).$
So, 484,024 is divisible by 8.

We circle the numbers that are divisible by 8.

4,828   ⟨8,064⟩   80,402   ⟨484,560⟩   ⟨484,024⟩

43. We look at the five numbers separately.

    We begin with 9,045.
    $$9,045 = 9,000+45$$
    $$= (9\times 1,000)+(9\times 5)$$
    $$= 9\times(1,000+5).$$
    So, 9,045 is divisible by 9.

    Then, we look at 3,610.
    $$3,610 = 3,600+10$$
    $$= 3,600+9+1$$
    $$= (9\times 400)+(9\times 1)+1$$
    $$= 9\times(400+1)+1.$$
    3,610 is 1 more than a multiple of 9.
    Therefore 3,610 is not divisible by 9.

    Next, we look at 45,726.
    $$45,726 = 45,000+720+6$$
    $$= (9\times 5,000)+(9\times 80)+6$$
    $$= 9\times(5,000+80)+6.$$
    45,726 is 6 more than a multiple of 9.
    Therefore 45,726 is not divisible by 9.

    Next, we look at 63,027.
    $$63,027 = 63,000+27$$
    $$= (9\times 7,000)+(9\times 3)$$
    $$= 9\times(7,000+3).$$
    So, 63,027 is divisible by 9.

    Finally, we look at 365,418.
    $$365,418 = 360,000+5,400+18$$
    $$= (9\times 40,000)+(9\times 600)+(9\times 2)$$
    $$= 9\times(40,000+600+2).$$
    So, 365,418 is divisible by 9.

    We circle the numbers that are divisible by 9.

    (9,045)   3,610   45,726   (63,027)   (365,418)

44. Since $99,999+999 = 9\times(11,111+111)$, this sum is divisible by 9. So, the remainder of $(99,999+999)\div 9$ is **0**.

45. Since $100 = 99+1 = (9\times 11)+1$, we know that 100 is 1 more than a multiple of 9. So, $100\div 9$ has remainder **1**.

46. We know that 10 is 1 more than 9. So,
    $$10+10+10+10+10+10+10+10$$
    $$= (9+1)+(9+1)+(9+1)+(9+1)+(9+1)+(9+1)+(9+1)+(9+1)$$
    $$= (9+9+9+9+9+9+9+9)+1+1+1+1+1+1+1+1$$
    $$= (9+9+9+9+9+9+9+9)+8.$$

    $10+10+10+10+10+10+10+10$ is 8 more than a multiple of 9. So, the remainder of $(10+10+10+10+10+10+10+10)\div 9$ is **8**.

47. Since $1,000 = 999+1 = (9\times 111)+1$, we know that 1,000 is 1 more than a multiple of 9. So,
    $$1,000+1,000+1,000+1,000$$
    $$= (999+1)+(999+1)+(999+1)+(999+1)$$
    $$= (999+999+999+999)+1+1+1+1$$
    $$= (999+999+999+999)+4.$$

    $1,000+1,000+1,000+1,000$ is 4 more than a multiple of 9. So, the remainder of $(1,000+1,000+1,000+1,000)\div 9$ is **4**.

48. Since $10,000 = 9,999+1 = (9\times 1,111)+1$, we know that 10,000 is 1 more than a multiple of 9. Also, as we've seen in previous problems, 1,000 and 100 are each 1 more than a multiple of 9. So,
    $$10,000+1,000+100 = (9,999+1)+(999+1)+(99+1)$$
    $$= (9,999+999+99)+1+1+1$$
    $$= (9,999+999+99)+3.$$

    $10,000+1,000+100$ is 3 more than a multiple of 9. The remainder of $(10,000+1,000+100)\div 9$ is **3**.

49. As we've seen in previous problems, 1,000 is 1 more than a multiple of 9. So,
    $5,000 = 1,000+1,000+1,000+1,000+1,000$ is
    $1+1+1+1+1 = 5$ more than a multiple of 9.

    The remainder of $5,000\div 9$ is **5**.

50. As we've seen in previous problems, 1,000 and 10 are each 1 more than a multiple of 9.

    So, 3,000 is 3 more than a multiple of 9, and 30 is 3 more than a multiple of 9.

    Therefore, $3,000+30$ is $3+3 = 6$ more than a multiple of 9. The remainder of $(3,000+30)\div 9$ is **6**.

51. As we've seen in previous problems, 10,000, 1,000, and 100 are each 1 more than a multiple of 9.

    So, we know that 10,000 is 1 more than a multiple of 9, 2,000 is 2 more than a multiple of 9, and 300 is 3 more than a multiple of 9.

    Therefore, $10,000+2,000+300$ is $1+2+3 = 6$ more than a multiple of 9. The remainder of $(10,000+2,000+300)\div 9$ is **6**.

52. As we've seen in previous problems, 10,000, 1,000, 100, and 10 are each 1 more than a multiple of 9.

    So, we know that 50,000 is 5 more than a multiple of 9,
    4,000 is 4 more than a multiple of 9,
    300 is 3 more than a multiple of 9,
    20 is 2 more than a multiple of 9, and
    1 is 1 more than a multiple of 9.

    Therefore, $50,000+4,000+300+20+1$ is $5+4+3+2+1 = 15$ more than a multiple of 9.

    A number that is 15 more than a multiple of 9 is $15-9 = 6$ more than the next multiple of 9. So, the remainder of $(50,000+4,000+300+20+1)\div 9$ is **6**.

53. If the sum of a number's digits is divisible by 3, then the number is divisible by 3. If the sum of a number's digits is not divisible by 3, then the number is not divisible by 3. So, we compute the sum of each number's digits.

Number	Sum of digits
702	$7+0+2 = 9$
1,689	$1+6+8+9 = 24$
8,213	$8+2+1+3 = 14$
14,695	$1+4+6+9+5 = 25$
198,664	$1+9+8+6+6+4 = 34$
594,231	$5+9+4+2+3+1 = 24$

The only sums in the table that are divisible by 3 are 9 and 24. Therefore, 702, 1,689, and 594,231 are divisible by 3, and the other numbers are not.

(702) (1,689) 8,213 14,695 198,664 (594,231)

54. If the sum of a number's digits is divisible by 9, then the number is divisible by 9. If the sum of a number's digits is not divisible by 9, then the number is not divisible by 9. So, we compute the sum of each number's digits.

Number	Sum of digits
504	5+0+4 = 9
3,152	3+1+5+2 = 11
2,853	2+8+5+3 = 18
16,572	1+6+5+7+2 = 21
374,184	3+7+4+1+8+4 = 27
482,527	4+8+2+5+2+7 = 28

The only sums above that are divisible by 9 are 9, 18, and 27. Therefore, 504, 2,853, and 374,184 are divisible by 9, and the other numbers are not.

(504) 3,152 (2,853) 16,572 (374,184) 482,527

55. Since $\boxed{4\,1\,A\,6}$ is divisible by 9, the sum of these four digits is divisible by 9.

$4+1+6 = 11$, and adding the remaining digit (A) must give us a total that is divisible by 9.

The only multiple of 9 that we can get by adding a digit (from 0 to 9) to 11 is $11 + \boxed{7} = 18$.

Therefore, $A = 7$, and our four-digit number is 4,1$\underline{7}$6.

56. For $\boxed{2\,0\,B\,0}$ to be divisible by 3, the sum of the number's four digits must be divisible by 3.

$2+0+0 = 2$, and adding the remaining digit (B) must give us a total that is divisible by 3.

There are a few multiples of 3 that we can get by adding a digit (from 0 to 9) to 2:

$2 + \boxed{1} = 3,$
$2 + \boxed{4} = 6,$ and
$2 + \boxed{7} = 9.$

Therefore, B could be **1, 4, or 7**.

57. First, we look at the different groups of three digits we could use from the four digits. One way to find these groups is to choose which digit we leave out.

Group of three	Digit left out
3, 5, 7	1
1, 5, 7	3
1, 3, 7	5
1, 3, 5	7

The divisibility test for 9 depends on the sum of the digits of a number, so we compute the sum of each group's digits. A number is only divisible by 9 if the sum of the number's digits is divisible by 9.

Group of three	Sum of digits
3, 5, 7	3+5+7 = 15
1, 5, 7	1+5+7 = 13
1, 3, 7	1+3+7 = 11
1, 3, 5	1+3+5 = 9

Of these four sums, only 9 is divisible by 9. So, we can only arrange the digits 1, 3, and 5 to make a multiple of 9. *Any* three-digit number that uses the digits 1, 3, and 5 is divisible by 9. So, the order of the digits does not matter. We count the number of arrangements of the digits 1, 3, and 5.

We have 3 choices for the hundreds digit. Once we've chosen the hundreds digit, we have 2 choices for the tens digit, which leaves the 1 remaining digit to be the ones digit. All together, we have $3 \times 2 \times 1 = \mathbf{6}$ ways to arrange these three digits into a three-digit multiple of 9.

— *or* —

We list all of the possible three-digit numbers that use the digits 1, 3, and 5. We organize our list by writing the numbers from least to greatest:

135   315   513
153   351   531

All together, we count **6** ways to arrange these three digits into a three-digit multiple of 9.

58. Adding the digits of a multiple of 9 gives us a sum that is a multiple of nine (0, 9, 18, 27, and so on). Since the sum of 1's and 2's cannot be 0, the sum of Grogg's digits must be *at least* 9.

Grogg makes his number with only 1's and 2's. If Grogg's number has 4 or fewer digits, the sum of the digits of his number is less than or equal to $2+2+2+2 = 8$.

So, the smallest number Grogg could write must have at least 5 digits.

We have $2+2+2+2+1 = 9$, which is a multiple of 9. So, every number consisting of four 2's and one 1 is a multiple of 9.

To get the smallest number from all possible arrangements of these digits, we use 1 as the first (ten-thousands) digit: 12,222.

Any five-digit number that begins with 2 will be greater than 12,222. Also, we cannot use more 1's without needing more digits in the number, and any number with 6 or more digits is greater than 12,222.

Therefore, **12,222** is the smallest number Grogg can write.

59. First, we look at the different groups of three digits we could use from the four digits. One way to find these groups is to choose which digit we leave out.

Group of three	Digit left out
4, 6, 8	2
2, 6, 8	4
2, 4, 8	6
2, 4, 6	8

The divisibility test for 3 depends on the sum of the digits of a number, so we compute the sum of each group's digits. A number is only divisible by 3 if the sum of the number's digits is divisible by 3.

Group of three	Sum of digits
4, 6, 8	4+6+8 = 18
2, 6, 8	2+6+8 = 16
2, 4, 8	2+4+8 = 14
2, 4, 6	2+4+6 = 12

Of these four sums, only 18 and 12 are divisible by 3. So, we can arrange 4, 6, and 8 to make a multiple of 3, and we can arrange 2, 4, and 6 to make a multiple of 3.

Any three-digit number made up of one of these two sets of digits is divisible by 3. So, the order of the digits does not matter. We count the number of ways we can arrange each set of digits.

For the group 4, 6, and 8, we have 3 choices for the hundreds digit. Once we've chosen the hundreds digit, we have 2 choices for the tens digit, which leaves the 1 remaining digit as the ones digit. So, there are $3\times 2\times 1 = 6$ ways to arrange these three digits into a three-digit multiple of 3.

Similarly, there are $3\times 2\times 1 = 6$ ways to arrange the digits 2, 4, and 6 into a three-digit multiple of 3.

All together, we can make $6+6 = $ **12** three-digit multiples of 3 by arranging three of the four digits shown.

— *or* —

We list all of the possible three-digit numbers that we can make with 4, 6, and 8. We organize our work by listing all of the numbers from least to greatest:

| 468 | 648 | 846 |
| 486 | 684 | 864 |

Then, we list all of the possible three-digit numbers that we can make with 2, 4, and 6:

| 246 | 426 | 624 |
| 264 | 462 | 642 |

All together, we count $6+6 = $ **12** three-digit multiples of 3.

60. If a number is even, its units digit is even (0, 2, 4, 6, or 8). Since $\boxed{S\,T\,U}$, $\boxed{T\,U\,S}$, and $\boxed{U\,S\,T}$ are all even, we know that all of the digits S, T, and U are even. So, the sum $S+T+U$ is even.

Since $\boxed{S\,T\,U}$, $\boxed{T\,U\,S}$, and $\boxed{U\,S\,T}$ are all divisible by 9, the sum $S+T+U$ is a multiple of 9. The sum of the digits of a three-digit number cannot be greater than $9+9+9 = 27$. So, the sum of the digits of a three-digit multiple of 9 can only be 9, 18, or 27. Of these, only 18 is even. Therefore, $S+T+U$ is **18**.

We do not need to find the individual values of S, T, or U.

FACTORS Divisibility Tests 18-19

61. Every multiple of 25 ends in 00, 25, 50, or 75. Since 4_7_ has a 7 in the tens place and is divisible by 25, we know that it must end in 7<u>5</u>. So, the ones digit is 5.

Now, we have 4_75. For 4_75 to be a multiple of 9, the sum of its digits must be a multiple of 9. The sum of the three known digits is $4+7+5 = 16$. The only multiple of 9 that we can get by adding a digit (from 0 to 9) to 16 is $16+\boxed{2} = 18$. Therefore, the hundreds digit is 2.

So, the four-digit number is **4275**.

62. Every multiple of 5 ends in 0 or 5. So, the units digit of 57_ _ is either 0 or 5. However, we also know that 57_ _ is divisible by 4. Multiples of 4 are even, so 57_ _ cannot end in 5. The units digit is 0.

Now, we have 57_0. For 57_0 to be divisible by 9, the sum of its digits must be a multiple of 9. The sum of the known digits is $5+7+0 = 12$. The only multiple of 9 that we can get by adding a digit (from 0 to 9) to 12 is $12+\boxed{6} = 18$. Therefore, the tens digit is 6.

The four-digit number is **5760**.

We can quickly check that the last two digits (60) form a two-digit number that is divisible by 4, so this number is definitely divisible by 4.

63. Since the sum of the digits $2+3+4+9 = 18$ is a multiple of 3, any arrangement of 2, 3, 4, and 9 will create a number that is divisible by 3.

Since the number must be even, we can only choose an even digit (2 or 4) for the units place.

We list the arrangements that end with 2, in order from least to greatest.

| 3,492 | 4,392 | 9,342 |
| 3,942 | 4,932 | 9,432 |

We also list the arrangements that end with 4, in order from least to greatest.

| 2,394 | 3,294 | 9,234 |
| 2,934 | 3,924 | 9,324 |

All together, we count $6+6 = $ **12** ways to arrange the four digits into an even number divisible by 3.

— *or* —

Since the sum of the digits $2+3+4+9 = 18$ is a multiple of three, any arrangement of 2, 3, 4, and 9 will create a number that is divisible by 3.

For a number to be even, its units digit must be even. From these four digits, we have 2 choices (either 2 or 4) for the units digit.

Once we have chosen the units digit, we have 3 choices for the tens digit. Then, we have 2 choices for the hundreds digit, leaving the 1 remaining digit for the thousands digit.

All together, there are $2\times 3\times 2\times 1 = $ **12** ways to arrange the four digits into an even number divisible by 3.

64. The sum of the digits of a multiple of 3 is always a multiple of 3. To use the digits 8 and 9 at least once each, we must make at least a two-digit number: 89 or 98. However, $8+9 = 17$ is not a multiple of 3, so neither 89 nor 98 is divisible by 3.

So, the number must contain at least three digits. To make a three-digit number using 8's and 9's, with at least one of each, we can use one or two 8's. However, neither $8+9+9 = 26$ nor $8+8+9 = 25$ is divisible by 3.

So, the number must contain at least four digits. To make a four-digit number from 8's and 9's, with at least one of each, we can use one, two, or three 8's. Of $8+9+9+9 = 35$, $8+8+9+9 = 34$, and $8+8+8+9 = 33$, only 33 is divisible by 3. So, our digits are 8, 8, 8, and 9.

To make our number as small as possible, we use the largest digit (9) as the units digit: 8,889. Placing the 9 in any other spot will give us a number greater than 8,889.

Therefore, **8,889** is the smallest multiple of 3 that uses only the digits 8 and 9, with at least one of each.

— *or* —

The sum of the digits of a multiple of 3 is always a multiple of 3. We need to find a sum of 8's and 9's that is divisible by 3. Since 9 is divisible by 3, subtracting 9 from any multiple of 3 gives another multiple of 3. So, a number's divisibility by 3 is not affected by any digits that are 9's.

Therefore, for a sum of 8's and 9's to be divisible by 3, the sum of the 8's must be divisible by 3. We know that 8 and 8+8 are not divisible by 3, but 8+8+8 = 24 is. So, we need at least three 8's to make a number divisible by 3.

Then, we are required to use at least one 9. To keep our number as small as possible, we use only one 9. So, our digits are 8, 8, 8, and 9.

To make our number as small as possible, we use the largest digit (9) as the units digit: 8,889. Placing the 9 in any other spot will give us a number greater than 8,889.

Therefore, **8,889** is the smallest multiple of 3 that uses only the digits 8 and 9, with at least one of each.

65. The smallest prime number is 2. The units digit of 736 is even, so 736 is divisible by 2.
2 is the smallest prime factor of 736.

66. We check for divisibility by primes in order, starting with 2. The units digit of 1,625 is not even, so 1,625 is not divisible by 2. The sum of the digits of 1,625 is 1+6+2+5 = 14, so 1,625 is not divisible by 3. The units digit of 1,625 is 5, so 1,625 is divisible by 5.
5 is the smallest prime factor of 1,625.

67. The units digit of 217 is not even, so 217 is not divisible by 2. The sum of the digits of 217 is 2+1+7 = 10, so 217 is not divisible by 3. The units digit of 217 is not 0 or 5, so 217 is not divisible by 5. The next-smallest prime is 7. Unfortunately, we don't have a nice divisibility test for 7. However, we can write
217 = 210+7 = (7×30)+(7×1) = 7×(30+1).
So, 217 is divisible by 7.
7 is the smallest prime factor of 217.

68. The units digit of 81,135 is not even, so 81,135 is not divisible by 2. The sum of the digits of 81,135 is 8+1+1+3+5 = 18, which is divisible by 3.
3 is the smallest prime factor of 81,135.

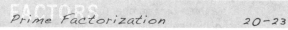

69. Since 110 = 10×$\boxed{11}$, we know that 11 is the prime that goes in the circle on the right.

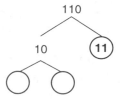

Next, we factor 10 into 2×5.

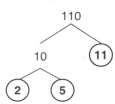

Since there are no composite numbers left to factor, we are finished! Writing these factors in order from least to greatest, the prime factorization of 110 is **2×5×11**.

70. Since 126 = 6×$\boxed{21}$, we know that 21 is the composite number that goes in the blank on the right.

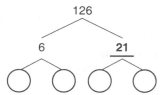

Next, we factor 6 into 2×3 and 21 into 3×7.

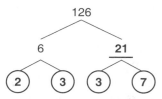

Since there are no composite numbers left to factor, we are finished! Writing these factors from least to greatest and using exponents for repeated factors, the prime factorization of 126 is **2×3×3×7 = 2×3² ×7**.

For each of the following problems, you may have factored the numbers with different steps to arrive at the same final prime factorization.

71.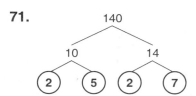

Writing these factors in order from least to greatest and using exponents for repeated factors, we have 140 = **2² ×5×7**.

72.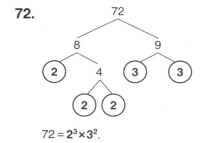

72 = **2³ ×3²**.

73.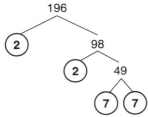

$196 = \mathbf{2^2 \times 7^2}$.

74.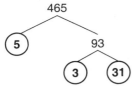

$465 = \mathbf{3 \times 5 \times 31}$.

75.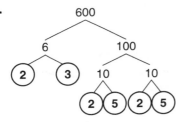

$600 = \mathbf{2^3 \times 3 \times 5^2}$.

76.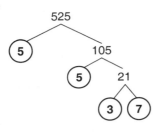

$525 = \mathbf{3 \times 5^2 \times 7}$.

77. $1{,}800 = 3 \times 600$. We previously found that the prime factorization of 600 is $2^3 \times 3 \times 5^2$.

We use this to find the prime factorization of 1,800.

$1{,}800 = 3 \times 600 = 3 \times (2^3 \times 3 \times 5^2)$
$= 3 \times 2 \times 2 \times 2 \times 3 \times 5 \times 5 = \mathbf{2^3 \times 3^2 \times 5^2}$.

78. $1{,}050 = 2 \times 525$. We previously found that the prime factorization of 525 is $3 \times 5^2 \times 7$.

We use this to find the prime factorization of 1,050.

$1{,}050 = 2 \times 525 = 2 \times (3 \times 5^2 \times 7)$
$= \mathbf{2 \times 3 \times 5^2 \times 7}$.

79. $6{,}000 = 10 \times 600$.

The prime factorization of 600 is $2^3 \times 3 \times 5^2$.
The prime factorization of 10 is 2×5.

We combine these to find the prime factorization of 6,000.

$6{,}000 = 10 \times 600 = (2 \times 5) \times (2^3 \times 3 \times 5^2)$
$= 2 \times 5 \times 2 \times 2 \times 2 \times 3 \times 5 \times 5 = \mathbf{2^4 \times 3 \times 5^3}$.

80. $5{,}250 = 10 \times 525$.
The prime factorization of 525 is $3 \times 5^2 \times 7$.
The prime factorization of 10 is 2×5.

We combine these to find the prime factorization of 5,250.

$5{,}250 = 10 \times 525 = (2 \times 5) \times (3 \times 5^2 \times 7)$
$= 2 \times 5 \times 3 \times 5 \times 5 \times 7 = \mathbf{2 \times 3 \times 5^3 \times 7}$.

81. $600 = 2 \times 300$.
The prime factorization of 600 is $2^3 \times 3 \times 5^2$.
We write out the factors and regroup to see that
$2 \times 2 \times 2 \times 3 \times 5 \times 5 = 2 \times (2 \times 2 \times 3 \times 5 \times 5) = 2 \times 300$.

So, the prime factorization of 300 is
$2 \times 2 \times 3 \times 5 \times 5 = \mathbf{2^2 \times 3 \times 5^2}$.

82. $525 = 5 \times 105$.
The prime factorization of 525 is $3 \times 5^2 \times 7$.

We write out the factors and regroup to see that
$3 \times 5 \times 5 \times 7 = 5 \times (3 \times 5 \times 7) = 5 \times 105$.

So, the prime factorization of 105 is $\mathbf{3 \times 5 \times 7}$.

83. We use division and our divisibility tests to look for prime factors of 87.

Since $8 + 7 = 15$ is divisible by 3, we know that 87 is divisible by 3. We find that $87 = 3 \times 29$. Both 3 and 29 are prime, so the prime factorization of 87 is $\mathbf{3 \times 29}$.

84. We use division and our divisibility tests to look for prime factors of 113.

113 is not even, so it is not divisible by 2.

$1 + 1 + 3 = 5$ is not divisible by 3, so we know that 113 is not divisible by 3.

113 does not end in 0 or 5, so we know that 113 is not divisible by 5.

We do not have a simple divisibility test for 7. To factor more quickly, we note that $7 \times 15 = 105$, so $7 \times 16 = 112$. Therefore, 113 is 1 more than a multiple of 7 and is not divisible by 7.

11 is the next prime. $11 \times 10 = 110$, so 113 is not divisible by 11.

Then, since $11 \times 11 = 121$, any number that is larger than 11 must be multiplied by a number that is smaller than 11 to get 113. So, we don't need to check any more primes.

The only factors of 113 are 1 and 113, so 113 is prime. The prime factorization of a prime is the number itself: **113**.

85. We use division and our divisibility tests to look for prime factors of 441.

Since $4 + 4 + 1 = 9$ is divisible by 3 and by 9, we know that 441 is divisible by 3 and by 9.

To factor more quickly, we note that $9 \times 50 = 450$, so $9 \times 49 = 450 - 9 = 441$. We begin our factor tree as shown.

Then, $9 = 3\times 3$ and $49 = 7\times 7$. Since 3 and 7 are both prime, we circle the 3's and 7's.

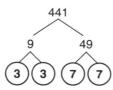

So, the prime factorization of 441 is $3\times 3\times 7\times 7 = \mathbf{3^2\times 7^2}$.

86. Since 910 ends in a 0, we know that 910 is divisible by 10. We have $910 = 10\times 91$.

Then, we factor 91 and 10. We have $10 = 2\times 5$ and $7\times 13 = 91$. Since 2, 5, 7, and 13 are all prime, we circle all four primes.

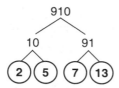

So, the prime factorization of 910 is $\mathbf{2\times 5\times 7\times 13}$.

87. Since 406 is even, we begin our factor tree with 2×203. Since 2 is prime, we circle the 2.

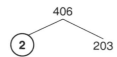

Then, we check to see if 203 is prime.

203 is not even, so it is not divisible by 2.

$2+0+3 = 5$ is not divisible by 3, so we know that 203 is not divisible by 3.

203 does not end in 0 or 5, so we know that 203 is not divisible by 5.

We do not have a simple divisibility test for 7. To factor more quickly, we note that $7\times 30 = 210$, so $7\times 29 = 210-7 = 203$. Therefore, 203 is divisible by 7. Since 7 and 29 are both prime, we circle 7 and 29.

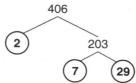

So, the prime factorization of 406 is $\mathbf{2\times 7\times 29}$.

88. We use division and our divisibility tests to look for prime factors of 357. Since $3+5+7 = 15$, we know that 357 is divisible by 3.

$3\times 120 = 360$, so $3\times 119 = 357$. We know that 3 is prime, so we circle it. We begin our factor tree as shown.

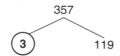

Then, we check to see if 119 is prime.

119 is not even, so it is not divisible by 2.

$1+1+9 = 11$ is not divisible by 3, so we know that 119 is not divisible by 3.

119 does not end in 0 or 5, so we know that 119 is not divisible by 5.

We do not have a simple divisibility test for 7. To factor more quickly, we note that $7\times 15 = 105$, so $7\times 16 = 112$ and $7\times 17 = 119$. Therefore, 119 is divisible by 7. Since 7 and 17 are both prime, we circle 7 and 17.

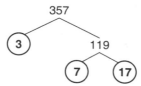

So, the prime factorization of 357 is $\mathbf{3\times 7\times 17}$.

Factor Blobs 24-25

We will use the first four solutions to illustrate four strategies you can use to help solve these puzzles.

89. **Strategy 1**: Look at the corners first.
A number in a corner only touches two adjacent numbers, and it must be grouped with at least one adjacent number.

For example, the 9 in the top left must be grouped with the 3 below it or the 5 to its right. Since $9\times 5 = 45$, we pair the 9 with the 5 as shown.

9	5	15
3	5	3
3	15	3

Then, the 15 in the top right must be paired with the 3 below it.

9	5	15
3	5	3
3	15	3

The remaining numbers are grouped into blobs as shown so that the product of the numbers in each blob is 45.

9	5	15
3	5	3
3	15	3

90. **Strategy 2**: Look at the bigger numbers.

For example, since $14\times 2 = 28$, the 14 can only be paired with the 2 to its right.

7	1	1
14	2	2
4	7	2

The remaining numbers are grouped into blobs as shown so that the product of the numbers in each blob is 28.

7	1	1
14	2	2
4	7	2

91. **Strategy 3**: Use prime factorizations.
Prime factorizations of both the target number and the composite numbers in the grid make it easier to see how to group the numbers.

For example, $750 = 2\times3\times5\times5\times5$. We also write the prime factorizations of the numbers in the grid.

So, we want to group one 2, one 3, and three 5's into each number blob.

The 25 in the top left has two 5's, and it needs to be grouped with an additional 2, 3, and 5. If we pair the 25 with the 15 to its right, we have one 3 and three 5's and still need a 2. But, there is no way to include a 2 in this blob. So, we group 25 with the 30 ($2\times3\times5$) below it to make a blob.

We use this strategy to group the remaining numbers into blobs as shown so that each blob has one 2, one 3, and three 5's.

92. <u>Strategy 4:</u> Separate numbers that cannot be in the same blob with "walls."

For example, since $63 = 3\times3\times7$, we need two 3's and one 7 in each number blob. We cannot have two 7's in the same blob. So, we draw a 'wall' between pairs of adjacent 7's. Similarly, we can separate the 21 from the 7.

We also cannot have more than two 3's in the same blob. So, we separate the two 9's.

We then group the remaining numbers into blobs as shown so that the product of the numbers in each blob is 63.

To solve the remaining Factor Blob puzzles, we use the strategies outlined in the previous solutions.

93. $42 = 2\times3\times7$.

94. $60 = 2^2\times3\times5$.

95. $140 = 2^2\times5\times7$.

96. $24 = 2^3\times3$.

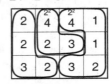

97. $900 = 2^2\times3^2\times5^2$.

98. $99 = 3^2\times11$.

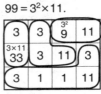

99. $96 = 2^5\times3$.

100. $72 = 2^3\times3^2$.

FACTORS
Pyramid Descent 26-28

101. $36 = 2^2\times3^2$. We find a path that gives us two 2's and two 3's.

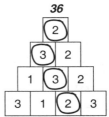

102. $24 = 2^3\times3$. We find a path that gives us three 2's and one 3.

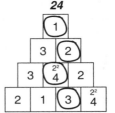

103. $84 = 2^2\times3\times7$. We find a path that gives us two 2's, one 3, and one 7.

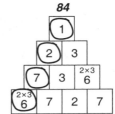

104. $120 = 2^3\times3\times5$. We find a path that gives us three 2's, one 3, and one 5.

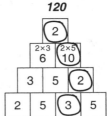

105. $180 = 2^2\times3^2\times5$.

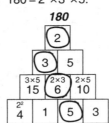

106. $216 = 2^3\times3^3$.

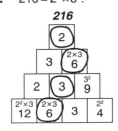

107. $210 = 2 \times 3 \times 5 \times 7$.

108. $999 = 3^3 \times 37$.

109. $72 = 2^3 \times 3^2$.

110. $400 = 2^4 \times 5^2$.

111. $1485 = 3^3 \times 5 \times 11$.

112. $720 = 2^4 \times 3^2 \times 5$.

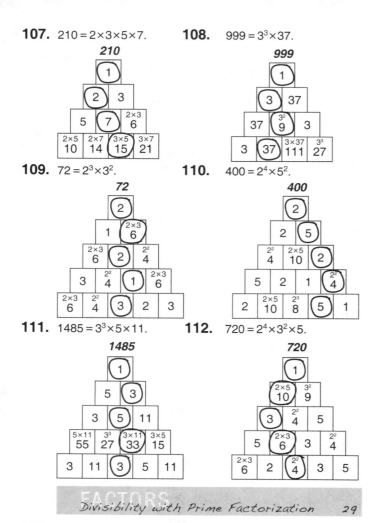

113. Instead of dividing 890,428 by each number, we consider the prime factorization of each number.

The prime factorization of 890,428 is $2^2 \times 7^3 \times 11 \times 59$.

First, we look at $4 = 2^2$. Since
$890{,}428 = \boxed{2^2} \times 7^3 \times 11 \times 59 = 4 \times 7^3 \times 11 \times 59$,
we know that 4 is a factor of 890,428.
Alternatively, we could use our divisibility rule for 4 to determine that 890,4$\underline{28}$ is divisible by 4.

Then, we look at $8 = 2^3$. The prime factorization of 8 has three 2's. However, the prime factorization of 890,428 has only two 2's: $\boxed{2} \times \boxed{2} \times 7 \times 7 \times 7 \times 11 \times 59$.
Therefore, 8 is not a factor of 890,428.

Next, we look at $14 = 2 \times 7$. We look for one 2 and one 7 in the prime factorization of 890,428:
$\boxed{2} \times 2 \times \boxed{7} \times 7 \times 7 \times 11 \times 59$. Rearranging and grouping 2×7, we have $890{,}428 = (2 \times 7) \times 2 \times 7 \times 7 \times 11 \times 59$
$= (14) \times 2 \times 7 \times 7 \times 11 \times 59$.
Therefore, 14 is a factor of 890,428.

Then, we look at $21 = 3 \times 7$. The prime factorization of 21 has one 3 and one 7. However, the prime factorization of $890{,}428 = 2^2 \times 7^3 \times 11 \times 59$ does not include any 3's.
Therefore, 21 is not a factor of 890,428.

Next, we look at $22 = 2 \times 11$. The prime factorization of 890,428 includes a 2 and an 11:
$890{,}428 = \boxed{2} \times 2 \times 7 \times 7 \times 7 \times \boxed{11} \times 59$.
So, 22 is a factor of 890,428.

Finally, we look at $236 = 2^2 \times 59$. The prime factorization of 890,428 includes two 2's and a 59:
$890{,}428 = \boxed{2} \times \boxed{2} \times 7 \times 7 \times 7 \times 11 \times \boxed{59}$.
Therefore, 236 is a factor of 890,428.

Of the numbers listed, only 4, 14, 22, and 236 are factors of 890,428.

④ 8 ⑭ 21 ㉒ ㉟

114. We use the prime factorization of each number to determine which have $1{,}323 = 3^3 \times 7^2$ as a factor. We look for at least three 3's and two 7's in the prime factorization of each choice.

First, we look at $7{,}938 = 2 \times 3^4 \times 7^2$, which includes three 3's and two 7's: $2 \times \boxed{3} \times \boxed{3} \times \boxed{3} \times 3 \times \boxed{7} \times \boxed{7}$.
Therefore, 1,323 is a factor of 7,938.

Then, we look at $37{,}044 = 2^2 \times 3^3 \times 7^3$, which includes three 3's and two 7's: $2 \times 2 \times \boxed{3} \times \boxed{3} \times \boxed{3} \times \boxed{7} \times \boxed{7} \times 7$.
Therefore, 1,323 is a factor of 37,044.

Next, we look at $79{,}233 = 3 \times 7^4 \times 11$, which only includes one 3. Therefore, 1,323 is not a factor of 79,233.

Finally, we look at $361{,}179 = 3^4 \times 7^3 \times 13$, which includes three 3's and two 7's: $\boxed{3} \times \boxed{3} \times \boxed{3} \times 3 \times \boxed{7} \times \boxed{7} \times 7 \times 13$.
Therefore 1,323 is a factor of 361,179.

Of the numbers listed, 1,323 is a factor of only 7,938, 37,044, and 361,179.

⟨$7{,}938 = 2 \times 3^4 \times 7^2$⟩   ⟨$37{,}044 = 2^2 \times 3^3 \times 7^3$⟩
$79{,}233 = 3 \times 7^4 \times 11$ ⟨$361{,}179 = 3^4 \times 7^3 \times 13$⟩

115. We start by listing 1, since 1 is a factor of every number.

Then, we look at the prime factorization of each whole number from 2 to 19. We use these prime factorizations to identify which numbers are factors of $480{,}200 = 2 \times 2 \times 2 \times 5 \times 5 \times 7 \times 7 \times 7 \times 7$.

$2 = 2$	$8 = 2^3$	$14 = 2 \times 7$
$3 = 3$	$9 = 3^2$	$15 = 3 \times 5$
$4 = 2^2$	$10 = 2 \times 5$	$16 = 2^4$
$5 = 5$	$11 = 11$	$17 = 17$
$6 = 2 \times 3$	$12 = 2^2 \times 3$	$18 = 2 \times 3^2$
$7 = 7$	$13 = 13$	$19 = 19$

For a number to be a factor of 480,200, all of its prime factors must appear in the prime factorization of 480,200. Also, each of its prime factors must appear in the prime factorization of 480,200 at least as many times as its exponent indicates.

For example, $10 = 2 \times 5$ is a factor of 480,200 because the prime factorization of 480,200 has both a 2 and a 5: $480{,}200 = \boxed{2} \times 2 \times 2 \times \boxed{5} \times 5 \times 7 \times 7 \times 7 \times 7$.

However, $6 = 2 \times 3$ is not a factor of 480,200 because the prime factorization of 480,200 does not have a 3. Similarly, $16 = 2^4$ is not a factor of 480,200 because the prime factorization of 480,200 does not have *four* 2's.

The numbers **1, 2, 4, 5, 7, 8, 10, and 14** are the factors of $480{,}200 = 2^3 \times 5^2 \times 7^4$ that are less than 20.

116. We can solve a division problem using multiplication. For example, to answer $54 \div 9$, we find the number that we can multiply by 9 to get 54. Since $9 \times \boxed{6} = 54$, we have $54 \div 9 = \boxed{6}$.

Similarly, to answer $42{,}250 \div 325$, we find the number that we can multiply by $325 = 5^2 \times 13$ get $42{,}250 = 2 \times 5^3 \times 13^2$.

To get $42{,}250 = 2 \times 5^3 \times 13^2$, we multiply $325 = 5^2 \times 13$ by one 2, one 5, and one 13:

$$5^2 \times 13 \times \boxed{2 \times 5 \times 13} = 2 \times 5^3 \times 13^2.$$

So, $325 \times \boxed{2 \times 5 \times 13} = 42{,}250$.

Therefore, $42{,}250 \div 325 = 2 \times 5 \times 13 = \mathbf{130}$.

117. We notice that the expressions given are not prime factorizations. $6 = 2 \times 3$ is a composite number.

So, we use the given factorizations to find the *prime* factorizations of each number:

$$\begin{aligned}34{,}992 &= 2^2 \times 3^5 \times 6^2 \\ &= 2^2 \times 3^5 \times 6 \times 6 \\ &= 2^2 \times 3^5 \times (2 \times 3) \times (2 \times 3) \\ &= (2^2 \times 2 \times 2) \times (3^5 \times 3 \times 3) \\ &= 2^4 \times 3^7.\end{aligned}$$

$$\begin{aligned}1{,}296 &= 6^4 \\ &= 6 \times 6 \times 6 \times 6 \\ &= (2 \times 3) \times (2 \times 3) \times (2 \times 3) \times (2 \times 3) \\ &= (2 \times 2 \times 2 \times 2) \times (3 \times 3 \times 3 \times 3) \\ &= 2^4 \times 3^4.\end{aligned}$$

Now, we see that the prime factorization of $34{,}992 = 2^4 \times 3^7$ includes the prime factorization of $1{,}296 = 2^4 \times 3^4$:

$$\begin{aligned}34{,}992 &= 2^4 \times 3^7 \\ &= \boxed{2} \times \boxed{2} \times \boxed{2} \times \boxed{2} \times \boxed{3} \times \boxed{3} \times \boxed{3} \times 3 \times 3 \times 3 \\ &= (2^4 \times 3^4) \times 3 \times 3 \times 3 \\ &= (1{,}296) \times 3 \times 3 \times 3.\end{aligned}$$

So, **34,992 is divisible by 1,296**.

— *or* —

We group pairs of 2's and 3's in the prime factorization of $34{,}992$ to make 6's:

$$\begin{aligned}34{,}992 &= 2^2 \times 3^5 \times 6^2 \\ &= (2 \times 2) \times (3 \times 3 \times 3 \times 3 \times 3) \times 6 \times 6 \\ &= (2 \times 3) \times (2 \times 3) \times (3 \times 3 \times 3) \times 6 \times 6 \\ &= 6 \times 6 \times 6 \times 6 \times (3 \times 3 \times 3) \\ &= 6^4 \times 3^3 \\ &= 1{,}296 \times 3^3.\end{aligned}$$

So, **34,992 is divisible by 1,296**.

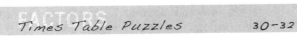

118. We label the factors A, B, C, and D as shown. A is a factor of both 14 and 35. D is a factor of both 35 and 15.

×	C	D
A	14	35
B		15

We look at the prime factorizations of 14, 35, and 15 to find numbers that are factors of both 14 and 35 and numbers that are factors of both 35 and 15.

$14 = 2 \times 7 \qquad 35 = 5 \times 7 \qquad 15 = 3 \times 5$

×	C	D
A	2×7 14	5×7 35
B		3×5 15

We know $A \times D = 35 = 5 \times 7$.

Since 7 is a factor of 35, but not 15, we know 7 is a factor of A and not of D.

×	C	D
7 A	2×7 14	5×7 35
B		3×5 15

Next, since 5 is a factor of 35, but not 14, we know 5 is a factor of D and not of A.

Since 35 does not have any additional prime factors, we know that A and D do not have any additional prime factors.

×	C	5 5
7 7	2×7 14	5×7 35
B		3×5 15

So, $A = 7$ and $D = 5$.

Then, $15 = B \times 5$. Since $15 = \boxed{3} \times 5$, we have $B = 3$.

Similarly, $14 = 7 \times C$. Since $14 = 7 \times \boxed{2}$, we have $C = 2$.

×	2 2	5 5
7 7	2×7 14	5×7 35
3 3	2×3 6	3×5 15

We multiply $B \times C = 3 \times 2 = 6$ to get the missing product.

Our times table is complete, and we check to make sure all of the products are correct.

To solve the next six puzzles, we use steps similar to those described above.

119. Step 1:

×		
2×2	2×2×3 12	2×2×5 20
	3×11 33	

Step 2:

×		3
2×2	2×2×3 12	2×2×5 20
	3×11 33	

Step 3:

×	3 3	5
2×2 4	2×2×3 12	2×2×5 20
11	3×11 33	

Final:

×	3 3	5 5
2×2 4	2×2×3 12	2×2×5 20
11 11	3×11 33	5×11 55

120. Step 1:

×		2×2
		2×2×3×7 84
	3×17 51	2×2×17 68

Step 2:

×		2×2
		2×2×3×7 84
17	3×17 51	2×2×17 68

Step 3:

×	3	2×2
3×7		2×2×3×7 84
17 17	3×17 51	2×2×17 68

Final:

×	3 3	2×2 4
3×7 21	3×3×7 63	2×2×3×7 84
17 17	3×17 51	2×2×17 68

121. Step 1, Step 2, Step 3, Final (factor tables as shown)

122. Step 1, Step 2, Step 3, Final (factor tables as shown)

123. Step 1, Step 2, Step 3, Final (factor tables as shown)

124. Step 1, Step 2, Step 3, Final (factor tables as shown)

125. We label the factors A, B, C, and D as shown. Since the prime factorization of 144 includes four 2's ($2 \times 2 \times 2 \times 2$), but the prime factorization of 270 includes only one 2, we know that the prime factorization of A includes at least three 2's.

×	C	D
A — $2\times2\times2$	$2\times2\times2\times3\times3$ — 144	$2\times2\times2\times5\times5$ — 200
B	$2\times3\times3\times3\times5$ — 270	

Since 144 and 200 do not share any additional prime factors, A does not have any additional prime factors. So, $A = 2\times2\times2 = 8$.

×	C	D
$2\times2\times2$ — 8	$2\times2\times2\times3\times3$ — 144	$2\times2\times2\times5\times5$ — 200
B	$2\times3\times3\times3\times5$ — 270	

Then, $144 = (2\times2\times2) \times C$. Since $144 = (2\times2\times2) \times \boxed{2\times3\times3}$, we have $C = 2\times3\times3 = 18$.

×	$2\times3\times3$ — 18	D
$2\times2\times2$ — 8	$2\times2\times2\times3\times3$ — 144	$2\times2\times2\times5\times5$ — 200
B	$2\times3\times3\times3\times5$ — 270	

Then, $270 = B \times (2\times3\times3)$. Since $270 = \boxed{3\times5} \times (2\times3\times3)$, we have $B = 3\times5 = 15$.

×	$2\times3\times3$ — 18	D
$2\times2\times2$ — 8	$2\times2\times2\times3\times3$ — 144	$2\times2\times2\times5\times5$ — 200
3×5 — 15	$2\times3\times3\times3\times5$ — 270	

Then, $200 = (2\times2\times2) \times D$. Since $200 = (2\times2\times2) \times \boxed{5\times5}$, we have $D = 5\times5 = 25$.

×	$2\times3\times3$ — 18	5×5 — 25
$2\times2\times2$ — 8	$2\times2\times2\times3\times3$ — 144	$2\times2\times2\times5\times5$ — 200
3×5 — 15	$2\times3\times3\times3\times5$ — 270	

We multiply $15 \times 25 = 375$ to get the missing product.

×	$2\times3\times3$ — 18	5×5 — 25
$2\times2\times2$ — 8	$2\times2\times2\times3\times3$ — 144	$2\times2\times2\times5\times5$ — 200
3×5 — 15	$2\times3\times3\times3\times5$ — 270	$3\times5\times5\times5$ — 375

126. We label the factors A, B, C, D, and E as shown. Since 3×3 is a factor of 72, but not of 32, we know that 3×3 is a factor of B.

×	C	D	E
A	$2\times2\times3\times5$ — 60	$2\times2\times2\times2\times2$ — 32	
3×3 — B		$2\times2\times2\times3\times3$ — 72	$3\times3\times7$ — 63

Since 7 is a factor of 63, but not of 72, we know that 7 is a factor of E. Since 63 does not have any additional prime factors, we know that B and E do not have any additional prime factors. So, $B = 3\times3 = 9$ and $E = 7$.

×	C	D	7 — 7
A	$2\times2\times3\times5$ — 60	$2\times2\times2\times2\times2$ — 32	
3×3 — 9		$2\times2\times2\times3\times3$ — 72	$3\times3\times7$ — 63

Then, $72 = 9 \times D$. Since $72 = 9 \times \boxed{8}$, we have $D = 8$.

×	C	2×2×2 8	7 7
A	2×2×3×5 60	2×2×2×2×2 32	
3×3 9		2×2×2×3×3 72	3×3×7 63

Then, $32 = A \times 8$. Since $32 = \boxed{4} \times 8$, we have $A = 4$.

×	C	2×2 8	7 7
2×2 4	2×2×3×5 60	2×2×2×2×2 32	
3×3 9		2×2×2×3×3 72	3×3×7 63

Then, $60 = 4 \times C$. Since $60 = 4 \times \boxed{15}$, we have $C = 15$.

×	3×5 15	2×2 8	7 7
2×2 4	2×2×3×5 60	2×2×2×2×2 32	
3×3 9		2×2×2×3×3 72	3×3×7 63

We multiply $4 \times 7 = 28$ and $9 \times 15 = 135$ to get the two missing products.

×	3×5 15	2×2×2 8	7 7
2×2 4	2×2×3×5 60	2×2×2×2×2 32	2×2×7 28
3×3 9	3×3×3×5 135	2×2×2×3×3 72	3×3×7 63

127. Since 17 is a factor of 68, but not of 88, we know that 17 is a factor of E.

×	C	D	17 E
A	2×2×2×11 88		2×2×17 68
B		5×5×7 175	7×17 119

Since 7 is a factor of 119, but not of 68, we know that 7 is a factor of B. Since 119 does not have any additional prime factors, we know that B and E do not have any additional prime factors. So, $B = 7$ and $E = 17$.

×	C	D	17 17
A	2×2×2×11 88		2×2×17 68
7 7		5×5×7 175	7×17 119

Then, $68 = A \times 17$. Since $68 = \boxed{4} \times 17$, we have $A = 4$.

×	C	D	17 17
2×2 4	2×2×2×11 88		2×2×17 68
7 7		5×5×7 175	7×17 119

Then, $88 = 4 \times C$. Since $88 = 4 \times \boxed{22}$, we have $C = 22$.

×	2×11 22	5×5 D	17 17
2×2 4	2×2×2×11 88		2×2×17 68
7 7		5×5×7 175	7×17 119

Then, $175 = 7 \times D$. Since $175 = 7 \times \boxed{25}$, we have $D = 25$.

×	2×11 22	5×5 25	17 17
2×2 4	2×2×2×11 88		2×2×17 68
7 7		5×5×7 175	7×17 119

We multiply $4 \times 25 = 100$ and $7 \times 22 = 154$ to get the two missing products.

×	2×11 22	5×5 25	17 17
2×2 4	2×2×2×11 88	2×2×5×5 100	2×2×17 68
7 7	2×7×11 154	5×5×7 175	7×17 119

128. Since 3×3 is a factor of 36, but not of 56, we know that 3×3 is a factor of B.

×	C	D	E
A	2×2×2×7 56	2×7×7 98	
3×3 B	2×2×3×3 36		2×3×3×5 90

Since the prime factorization of 56 includes three 2's (2×2×2), but the prime factorization of 98 includes only one 2, we know that the prime factorization of C includes *at least* two 2's.

Since 36 does not have any additional prime factors, we know that B and C do not have any additional prime factors. So, $B = 3 \times 3 = 9$ and $C = 2 \times 2 = 4$.

×	2×2 4	D	E
A	2×2×2×7 56	2×7×7 98	
3×3 9	2×2×3×3 36		2×3×3×5 90

Then, $56 = A \times 4$. Since $56 = \boxed{14} \times 4$, we have $A = 14$.

×	2×2 4	D	E
2×7 14	2×2×2×7 56	2×7×7 98	
3×3 9	2×2×3×3 36		2×3×3×5 90

Then, $98 = 14 \times D$. Since $98 = 14 \times \boxed{7}$, we have $D = 7$. Similarly, $90 = 9 \times E$. Since $90 = 9 \times \boxed{10}$, we have $E = 10$.

×	2×2 4	7 7	2×5 10
2×7 14	2×2×2×7 56	2×7×7 98	
3×3 9	2×2×3×3 36		2×3×3×5 90

We multiply $14 \times 10 = 140$ and $9 \times 7 = 63$ to get the two missing products.

×	2×2 4	7 7	2×5 10
2×7 14	2×2×2×7 56	2×7×7 98	2×2×5×7 140
3×3 9	2×2×3×3 36	3×3×7 63	2×3×3×5 90

129. Since the prime factorization of 112 includes four 2's (2×2×2×2), but the prime factorization of 294 includes only one 2, we know that the prime factorization of B includes at least three 2's.

×	C	D	E
A	2×3×7×7 294	2×3×5×7 210	
2×2×2 B	2×2×2×7 112		2×2×2×11 88

Then, since 112 and 88 do not share any additional prime factors, B does not have any additional prime factors. So, $B = 2×2×2 = 8$.

×	C	D	E
A	2×3×7×7 294	2×3×5×7 210	
2×2×2 **8**	2×2×2×7 112		2×2×2×11 88

Then, $112 = 8 \times C$. Since $112 = (2 \times 2 \times 2) \times \boxed{2 \times 7} = 8 \times \boxed{14}$, we have $C = 14$.

Then, $88 = 8 \times E$. Since $88 = 8 \times \boxed{11}$, we have $E = 11$.

×	2×7 **14**	D	11 **11**
A	2×3×7×7 294	2×3×5×7 210	
2×2×2 **8**	2×2×2×7 112		2×2×2×11 88

Then, $294 = A \times 14$. Since $294 = \boxed{3 \times 7} \times (2 \times 7) = \boxed{21} \times 14$, we have $A = 21$.

×	2×7 **14**	D	11 **11**
3×7 **21**	2×3×7×7 294	2×3×5×7 210	
2×2×2 **8**	2×2×2×7 112		2×2×2×11 88

Then, $210 = 21 \times D$. Since $210 = 21 \times 10$, we have $D = 10$.

×	2×7 **14**	2×5 **10**	11 **11**
3×7 **21**	2×3×7×7 294	2×3×5×7 210	
2×2×2 **8**	2×2×2×7 112		2×2×2×11 88

We multiply $21 \times 11 = 231$ and $8 \times 10 = 80$ to get the two missing products.

×	2×7 **14**	2×5 **10**	11 **11**
3×7 **21**	2×3×7×7 294	2×3×5×7 210	3×7×11 **231**
2×2×2 **8**	2×2×2×7 112	2×2×2×5 **80**	2×2×2×11 88

130. Every number has itself as a factor, but the result of subtracting a number from itself is 0 (a loss). So, we only consider subtracting other factors.

If a game begins with the number 4, there are only two options for Player 1: subtract 1 or subtract 2.

If Player 1 subtracts 1 from 4, Player 1 writes 3:

⨯ 3

Player 2 must subtract 1 from 3 and writes 2:

⨯⨯ 2

Player 1 must subtract 1 from 2 and writes 1:

⨯⨯⨯ 1

Player 2 is forced to subtract 1 from 1 and writes 0.

⨯⨯⨯⨯ 0

So, it is better to play first in a game of Factor Nim that begins with the number 4, because Player 1 can guarantee that they will win by subtracting 1 from 4.

If, however, Player 1 begins by subtracting 2 from 4, Player 1 writes 2: ⨯ 2

Player 2 must subtract 1 from 2 and writes 1:

⨯⨯ 1

Player 1 must subtract 1 from 1 and writes 0:

⨯⨯⨯ 0

This is not a win. So, **Player 1 *must* begin by subtracting 1 to guarantee a win.**

131. We begin by listing the factors of 45: 1, 3, 5, 9, 15, 45. The results of subtracting a factor of 45 from 45 are

$45 - 1 = \textbf{44}$,
$45 - 3 = \textbf{42}$,
$45 - 5 = \textbf{40}$,
$45 - 9 = \textbf{36}$,
$45 - 15 = \textbf{30}$, and
$45 - 45 = \textbf{0}$.

None of these results are odd.

132. An odd number has no 2's in its prime factorization. So, no factor of an odd number can have a prime factor of 2. Therefore, all of the factors of an odd number are odd. Subtracting an odd number from another odd number will always give an even result. For example, $7 - 1 = 6$ and $45 - 9 = 36$.

Therefore, when we subtract an odd number's factor (which must be odd) from that odd number, the result is always even. **It is not possible to subtract a factor from an odd number to get an odd result.**

133. If a game begins with any even number, Player 1 can subtract 1 and write an odd number. For example, at the start of a game that begins with 60, Player 1 can subtract 1 and write 59.

In the previous problem, we discovered that subtracting a factor from an odd number always gives an even result. So, whenever Player 1 writes an odd number, Player 2 has no choice but to subtract an odd factor and write an even number. In the example game from the previous paragraph, Player 2 is given 59 and can only subtract 1 to write $59 - 1 = 58$.

From this even number, Player 1 can again subtract 1 (or any other available odd factor) to write an odd number.

Using this strategy, every number that Player 1 writes will be odd, while every number that Player 2 writes will be even. This continues until Player 2 writes 2, Player 1 writes 1, and Player 2 writes 0.

Player 1 can guarantee a win in a game of Factor Nim that begins at 60 (or any other even number) by always subtracting 1 (or any other available odd factor) and writing only odd numbers, forcing Player 2 to write only even numbers.

FACTORS Challenge Problems 34-37

134. If 4,878 is divisible by 3, then Xavier's original number is divisible by 3. If 4,878 is divisible by 9, then Xavier's original number is divisible by 9.

We find the sum of the digits of 4,878 to see if it is divisible by 3 or by 9.

$4+8+7+8 = 27$. Since 27 is divisible by 3 and by 9, we know that 4,878 is divisible by both 3 and 9. So, Xavier's number is divisible by **both 3 and 9**.

135. We check if this number is divisible by a number for which we have a divisibility test.

The units digit of the number is 1, so the number is not divisible by 2 or by 5.

The sum of the digits is $8+7+6+5+4+3+2+1 = 36$, which is a multiple of 3 and of 9. So, our original number is divisible by 3 and by 9.

A prime is only divisible by 1 and itself. Since 87,654,321 is divisible by both 3 and 9, we know that 87,654,321 is not prime.

136. First, we write 7! as a product: $7! = 7\times6\times5\times4\times3\times2\times1$.

Since 1 is not prime, we do not write 1 in the prime factorization of a number.

Both 4 and 6 are composite numbers.
The prime factorization of 4 is 2×2, and the prime factorization of 6 is 2×3.

Replacing each of these factors in the product with their prime factorizations, we get
$$7! = 7\times6\times5\times4\times3\times2\times1$$
$$= 7\times(2\times3)\times5\times(2\times2)\times3\times2.$$

Now, every factor in this product is prime. We reorder these factors from least to greatest and write the prime factorization using exponents.

$$7! = 7\times6\times5\times4\times3\times2\times1$$
$$= 7\times(2\times3)\times5\times(2\times2)\times3\times2$$
$$= 2\times2\times2\times2\times3\times3\times5\times7$$
$$= \mathbf{2^4\times3^2\times5\times7}.$$

137. The four smallest prime numbers are 2, 3, 5, and 7, and their product is $2\times3\times5\times7 = 210$.

Any number with additional prime factors and any number with larger prime factors is larger than 210. Therefore, **210** is the smallest composite number that has four different prime factors.

138. No even number between 4,020 and 4,030 is prime. So, we list the odd numbers between 4,020 and 4,030.

4,021   4,023   4,025   4,027   4,029

4,023 has digit sum $4+0+2+3 = 9$ and is therefore divisible by 3 (and 9). So, 4,023 is not prime.

4,029 is 6 more than 4,023, a multiple of 3. Since a number that is 6 more than a multiple of 3 is also a multiple of 3, we know that 4,029 is also divisible by 3. So, 4,029 is not prime.

Then, 4,025 has units digit 5 and is therefore divisible by 5. So, 4,025 is not prime.

4,021   ~~4,023~~   ~~4,025~~   4,027   ~~4,029~~

This leaves two numbers, 4,021 and 4,027. We are told that there are two primes between 4,020 and 4,030, so those primes must be **4,021 and 4,027**.

139. If a number is divisible by 2 and by 3, then both 2 and 3 will appear in the number's prime factorization: $2\times3\times$(something).

We can group the 2 and 3 to make $2\times3\times$(something) $= 6\times$(something).

Therefore, Cammie's test is correct. Any number that is divisible by both 2 and 3 must be divisible by 6.

140. Some numbers that are divisible by both 2 and 6 are divisible by 12, such as 24, 36, and 60.

However, other numbers such as 6, 18, and 30 are divisible by both 2 and 6, but are not divisible by 12.

So, Ralph's rule is not correct.

A number is divisible by 2 if it has at least one 2 in its prime factorization. A number is divisible by $6 = 2\times3$ if it has at least one 2 and one 3 in its prime factorization. For a number to be divisible by $12 = 2^2\times3$, it must have at least *two* 2's and one 3 in its prime factorization. So, any number that has *exactly* one 2 and one 3 in its prime factorization is divisible by 2 and by 6, but not by 12.

In general, if a number n is divisible by two numbers that share a prime factor, then n is not necessarily divisible by the product of those two numbers. For example, 10 and 25 share a prime factor (5). We can see that 50 is divisible by both 10 and 25, but not by $10\times25 = 250$.

*However, if a number n is divisible by two numbers that **do not** have any prime factors in common, then n **is** divisible by the product of those two numbers. For example, $14 = 2\times7$ and $15 = 3\times5$ do not have any prime factors in common. Any number that is divisible by both 14 and 15 is divisible by $14\times15 = 210$.*

141. Since our desired number is not divisible by 2, 3, or 5, 7 is the smallest prime that can divide our number.

Our number is composite, so we know that it must have at least one other prime factor.

As before, since our number is not divisible by 2, 3, or 5, 7 is the smallest prime we can multiply by 7 to make our number composite.

Therefore, $7\times7 = \mathbf{49}$ is the smallest composite number that is not divisible by 2, 3, or 5.

142. Consider the prime factorization of a number that is divisible by 2, 3, 4, 5, 6, and 9.

For a number to be divisible by 2, it must have at least one 2 in its prime factorization.

For a number to be divisible by 3, it must have at least one 3 in its prime factorization.

For a number to be divisible by 4 = 2×2, it must have at least two 2's in its prime factorization.

For a number to be divisible by 5, it must have at least one 5 in its prime factorization.

For a number to be divisible by 6 = 2×3, it must have at least one 2 and one 3 in its prime factorization.

For a number to be divisible by 9 = 3×3, it must have at least two 3's in its prime factorization.

Reviewing all of these, we see that any number that is divisible by 2, 3, 4, 5, 6, and 9 has at least two 2's, two 3's, and one 5 in its prime factorization.

So, the smallest number that is divisible by 2, 3, 4, 5, 6, and 9 is 2×2×3×3×5 = **180**.

Any multiple of 180 is also divisible by all of our desired numbers. For example, the four numbers below are all divisible by 2, 3, 4, 5, 6, and 9:

$2 \times 180 = 360 = 2 \times (2 \times 2 \times 3 \times 3 \times 5) = 2^3 \times 3^2 \times 5$,
$3 \times 180 = 540 = 3 \times (2 \times 2 \times 3 \times 3 \times 5) = 2^2 \times 3^3 \times 5$,
$4 \times 180 = 720 = 2 \times 2 \times (2 \times 2 \times 3 \times 3 \times 5) = 2^4 \times 3^2 \times 5$, and
$5 \times 180 = 900 = 5 \times (2 \times 2 \times 3 \times 3 \times 5) = 2^2 \times 3^2 \times 5^2$.

The five three-digit numbers that are divisible by 2, 3, 4, 5, 6 and 9 are **180, 360, 540, 720, and 900**. You may have named any one of these five!

143. An odd number cannot have a factor of 2. The next three primes are 3, 5, and 7, and their product is 3×5×7 = 105.

Any number with larger prime factors is larger than 105. Therefore, **105** is the smallest odd number that has three different prime factors.

144. We consider the prime factorization of both numbers: $16 = 2^4$ and $28 = 2^2 \times 7$.

The prime factorization of 16 contains four 2's. The prime factorization of 28 contains two 2's and one 7.

So, the prime factorization of any nonzero number that has both 16 and 28 as factors must contain at least four 2's and at least one 7.

The smallest such number must contain *only* these factors and no others: $2 \times 2 \times 2 \times 2 \times 7 = 2^4 \times 7 =$ **112**.

We check that 16 and 28 are both factors of 112:
$(2 \times 2 \times 2 \times 2) \times 7 = \underline{16} \times 7 = 112$, and
$(2 \times 2) \times (2 \times 2 \times 7) = 4 \times \underline{28} = 112$.

145. We consider the prime factorization of both numbers: $198 = 2 \times 3 \times 3 \times 11$ and $330 = 2 \times 3 \times 5 \times 11$.

Each of these numbers has one 2, one 3, and one 11 in its prime factorization. These three factors are the only common prime factors of 198 and 330. Their product is 2×3×11 = 66, a factor of both 198 and 330.

If we multiply 2×3×11 = 66 by any more prime factors, the result will *not* be a factor of *both* 198 and 330. For example, (2×3×11)×3 is a factor of 198 but not 330, and (2×3×11)×2 is not a factor of either 198 or 330.

So, the largest factor of both 198 and 330 is 2×3×11 = **66**.

146. We consider the list of numbers and cross out those that are not prime until only two are left. We start by crossing out the numbers that end in 2 or in 5. Numbers that end in 2 are even and therefore divisible by 2. Numbers that end in 5 are divisible by 5.

~~235~~   ~~325~~   523   723
237   327   527   ~~725~~
253   ~~352~~   ~~532~~ 732
257   357   537   ~~735~~
273   ~~372~~   ~~572~~ 752
~~275~~   ~~375~~   573   753

Next, we add the digits of each remaining number to use our divisibility test for 3. We notice that 2+3+7 = 12, which a multiple of 3. So, any arrangement of the digits 2, 3, and 7 is divisible by 3. Similarly, 3+5+7 = 15, so any arrangement of 3, 5, and 7 is divisible by 3. Among our remaining numbers, we cross out 237, 273, 327, 357, 537, 573, 723, and 753.

~~235~~   ~~325~~   523   ~~723~~
~~237~~   ~~327~~   527   ~~725~~
253   ~~352~~   ~~532~~ 732
257   ~~357~~   ~~537~~ ~~735~~
~~273~~   ~~372~~   ~~572~~ 752
~~275~~   ~~375~~   ~~573~~ ~~753~~

Finally, we look at the problem statement, which tells us that 253 = 11×23, and 527 = 17×31. We cross these out.

~~235~~   ~~325~~   (523) ~~723~~
~~237~~   ~~327~~   ~~527~~ ~~725~~
~~253~~   ~~352~~   ~~532~~ 732
(257)   ~~357~~   ~~537~~ ~~735~~
~~273~~   ~~372~~   ~~572~~ 752
~~275~~   ~~375~~   ~~573~~ ~~753~~

This leaves two numbers, which are the only three-digit primes with three different prime digits: **257 and 523**.

147. The product of Cora's, Dora's, and Flora's ages is 36. So, each sister's age must be a factor of 36: 1, 2, 3, 4, 6, 9, 12, 18, or 36.

We create a list of the possible ages of the three sisters. To organize our list, we start with the largest possible age for the oldest sister, 36.

If the oldest sister is 36, the product of the ages of the other sisters is 1. So, if the oldest sister is 36, the other two sisters must both be 1.

If the oldest sister is 18, the product of the ages of the other sisters is 2. So, if the oldest sister is 18, the other two sisters are 2 and 1.

If the oldest sister is 12, the product of the ages of the other sisters is 3. So, if the oldest sister is 12, the other two sisters are 3 and 1.

Age 1	Age 2	Age 3
36	1	1
18	2	1
12	3	1

If the oldest sister is 9, the product of the ages of the other sisters is 4. There are two ways for the product of two ages to be 4. They may be 4 and 1, or 2 and 2.

Age 1	Age 2	Age 3
36	1	1
18	2	1
12	3	1
9	4	1
9	2	2

If the oldest sister is 6, the product of the ages of the other sisters is 6. There are two ways for the product of two ages to be 6. They may be 6 and 1, or 3 and 2.

If the oldest sister is 4, the product of the ages of the other sisters is 9. Since the two other sisters are the same age as or younger than the oldest sister, the other two sisters can only be 3 and 3.

The oldest sister cannot be 3. If one sister is 3, at least one of the other sisters must be older than 3. We've already considered all those possibilities above.

So, our chart is complete.

Age 1	Age 2	Age 3
36	1	1
18	2	1
12	3	1
9	4	1
9	2	2
6	6	1
6	3	2
4	3	3

When Winnie looks at Lucas's locker number to see the sum of the sisters' ages, Winnie does not have enough information to know their ages.

We find the sums of all eight possibilities in our table:

Ages			Sum
36	1	1	38
18	2	1	21
12	3	1	16
9	4	1	14
9	2	2	13
6	6	1	13
6	3	2	11
4	3	3	10

13 is the only sum that appears more than once, so **13** must be Lucas's locker number.

If Lucas had locker 38, 21, or any of the other possible sums, Winnie would have been able to figure out his sisters' ages from the locker number.

Chapter 8: Solutions

FRACTIONS
Review — pages 39-43

1. The tick marks split the number line between 0 and 1 into 5 equal pieces. Each piece has length $\frac{1}{5}$. We count 3 lengths of $\frac{1}{5}$ from 0 and mark $\frac{3}{5}$ at the end of the third piece.

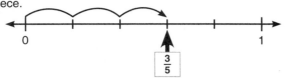

2. The tick marks split the number line between 0 and 1 into 8 equal pieces. Each piece has length $\frac{1}{8}$. We count 3 lengths of $\frac{1}{8}$ from 0 and mark $\frac{3}{8}$ at the end of the third piece.

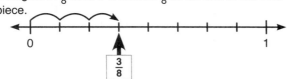

3. The tick marks split the number line between 0 and 1 into 9 equal pieces. Each piece has length $\frac{1}{9}$. We count 5 lengths of $\frac{1}{9}$ from 0 and mark $\frac{5}{9}$ at the end of the fifth piece.

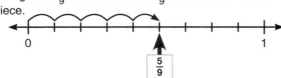

4. The tick marks split the number line between 0 and 1 into 11 equal pieces. Each piece has length $\frac{1}{11}$. We count 8 lengths of $\frac{1}{11}$ from 0 and mark $\frac{8}{11}$ at the end of the eighth piece.

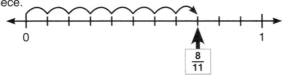

5. The tick marks split the number line between 0 and 1 into 4 equal pieces. Each piece has length $\frac{1}{4}$. We count 1 length of $\frac{1}{4}$ from 0 and mark $\frac{1}{4}$ at the end of the first piece.

6. The tick marks split the number line between 0 and 1 into 6 equal pieces. Each piece has length $\frac{1}{6}$. We count 5 lengths of $\frac{1}{6}$ from 0 and mark $\frac{5}{6}$ at the end of the fifth piece.

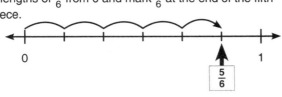

7. $\frac{48}{8}$ means $48 \div 8$. So, $\frac{48}{8} = 48 \div 8 = \mathbf{6}$.

8. $\frac{91}{7} = 91 \div 7 = \mathbf{13}$.

9. $\frac{55}{11} = 55 \div 11 = \mathbf{5}$.

10. $\frac{75}{15} = 75 \div 15 = \mathbf{5}$.

11. $\frac{156}{3} = 156 \div 3 = \mathbf{52}$.

12. $\frac{52}{52} = 52 \div 52 = \mathbf{1}$.

13. The missing numerator is the number that we divide by 7 to get 2. Since $2 \times 7 = 14$, we have $\boxed{14} \div 7 = 2$. So, $\frac{14}{7} = 2$.

14. Since $12 \times 3 = 36$, we have $\boxed{36} \div 3 = 12$. So, $\frac{36}{3} = \mathbf{12}$.

15. Since $17 \times 3 = 51$, we have $\boxed{51} \div 3 = 17$. So, $\frac{51}{3} = \mathbf{17}$.

16. Since $1 \times 5 = 5$, we have $\boxed{5} \div 5 = 1$. So, $\frac{5}{5} = \mathbf{1}$.

17. Since $11 \times 13 = 143$, we have $\boxed{143} \div 13 = 11$. So, $\frac{143}{13} = \mathbf{11}$.

18. Since $7 \times 15 = 105$, we have $\boxed{105} \div 15 = 7$. So, $\frac{105}{15} = \mathbf{7}$.

19. We know $\frac{16}{3}$ is between $\frac{15}{3} = 5$ and $\frac{18}{3} = 6$. $\frac{16}{3}$ is one third more than $\frac{15}{3} = 5$. So, written as a mixed number, $\frac{16}{3} = 5 + \frac{1}{3} = 5\frac{1}{3}$.

    We count one third past 5 to reach $5\frac{1}{3} = \frac{16}{3}$.

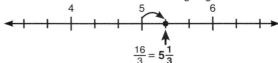

20. We know $\frac{55}{8}$ is between $\frac{48}{8} = 6$ and $\frac{56}{8} = 7$. $\frac{55}{8}$ is seven eighths more than $\frac{48}{8} = 6$. So, written as a mixed number, $\frac{55}{8} = 6 + \frac{7}{8} = 6\frac{7}{8}$.

    We count seven eighths past 6 to reach $6\frac{7}{8} = \frac{55}{8}$.

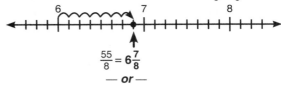

— or —

To locate $\frac{55}{8} = 6\frac{7}{8}$ on the given number line, we see that $\frac{55}{8}$ is one eighth less than $\frac{56}{8} = 7$.

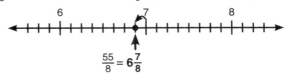

21. We know $\frac{39}{4}$ is between $\frac{36}{4} = 9$ and $\frac{40}{4} = 10$.

$\frac{39}{4}$ is three fourths more than $\frac{36}{4} = 9$. So, written as a mixed number, $\frac{39}{4} = 9 + \frac{3}{4} = 9\frac{3}{4}$.

We count three fourths past 9 to reach $9\frac{3}{4} = \frac{39}{4}$.

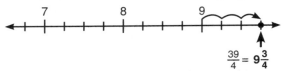

$\frac{39}{4} = 9\frac{3}{4}$

22. We know $\frac{34}{5}$ is between $\frac{30}{5} = 6$ and $\frac{35}{5} = 7$.

$\frac{34}{5}$ is four fifths more than $\frac{30}{5} = 6$. So, written as a mixed number, $\frac{34}{5} = 6 + \frac{4}{5} = 6\frac{4}{5}$.

To locate $\frac{34}{5} = 6\frac{4}{5}$ on the given number line, we see that $\frac{34}{5}$ is one fifth less than $\frac{35}{5} = 7$.

$\frac{34}{5} = 6\frac{4}{5}$

23. $9 \times \boxed{3} = 27$. So, to make an equivalent fraction with denominator 27, we multiply both the numerator and denominator of $\frac{7}{9}$ by 3.

$$\frac{7}{9} \xrightarrow{\times 3} \frac{21}{27}$$

24. $12 \div \boxed{6} = 2$. So, to make an equivalent fraction with numerator 2, we divide both the numerator and denominator of $\frac{12}{42}$ by 6.

$$\frac{12}{42} \xrightarrow{\div 6} \frac{2}{7}$$

25. $40 \div \boxed{4} = 10$. So, to make an equivalent fraction with denominator 10, we divide both the numerator and denominator of $\frac{56}{40}$ by 4.

$$\frac{56}{40} \xrightarrow{\div 4} \frac{14}{10}$$

26. $6 \times \boxed{12} = 72$. So, to make an equivalent fraction with numerator 72, we multiply both the numerator and denominator of $\frac{6}{5}$ by 12.

$$\frac{6}{5} \xrightarrow{\times 12} \frac{72}{60}$$

27. We begin with the first pair of fractions: $\frac{28}{42} = \frac{2}{-}$.

$28 \div \boxed{14} = 2$. So, to make an equivalent fraction with numerator 2, we divide both the numerator and denominator of $\frac{28}{42}$ by 14.

$$\frac{28}{42} \xrightarrow{\div 14} \frac{2}{3}$$

We look at the second pair of fractions: $\frac{2}{3} = \frac{-}{9}$.

$3 \times \boxed{3} = 9$. So, to make an equivalent fraction with denominator 9, we multiply both the numerator and denominator of $\frac{2}{3}$ by 3.

$$\frac{2}{3} \xrightarrow{\times 3} \frac{6}{9}$$

Finally, we look at the third pair of fractions: $\frac{6}{9} = \frac{18}{-}$.

$6 \times \boxed{3} = 18$. So, to make an equivalent fraction with numerator 18, we multiply both the numerator and denominator of $\frac{6}{9}$ by 3.

$$\frac{6}{9} \xrightarrow{\times 3} \frac{18}{27}$$

All together, we have $\frac{28}{42} = \frac{2}{3} = \frac{6}{9} = \frac{18}{27}$.

28. Rather than trying to find the number that we can multiply or divide 30 by to get 20, we notice that the fraction on the left side is not in simplest form.

We write an equivalent fraction in simplest form by dividing both 24 and 30 by 6.

$$\frac{24}{30} \xrightarrow{\div 6} \frac{4}{5}$$

Now, we use this new fraction to find an equivalent fraction with denominator 20.

$5 \times \boxed{4} = 20$. So, to make an equivalent fraction with denominator 20, we multiply both the numerator and denominator of $\frac{4}{5}$ by 4.

$$\frac{4}{5} \xrightarrow{\times 4} \frac{16}{20}$$

So, we have $\frac{24}{30} = \frac{16}{20}$.

29. *Three* fifths is less than *four* fifths. So, $\frac{3}{5}$ $\boxed{<}$ $\frac{4}{5}$.

30. Eighths are larger than ninths. So, $\frac{3}{8}$ $\boxed{>}$ $\frac{3}{9}$.

31. We compare $\frac{3}{4}$ to $\frac{5}{8}$ by converting $\frac{3}{4}$ to an equivalent fraction with denominator 8.

$$\frac{3}{4} \xrightarrow{\times 2} \frac{6}{8}$$

$\frac{6}{8} > \frac{5}{8}$, so $\frac{3}{4}$ $\boxed{>}$ $\frac{5}{8}$.

32. We compare $\frac{3}{5}$ to $\frac{11}{15}$ by converting $\frac{3}{5}$ to an equivalent fraction with denominator 15.

$$\frac{3}{5} \xrightarrow{\times 3} \frac{9}{15}$$

$\frac{9}{15} < \frac{11}{15}$, so $\frac{3}{5}$ $\boxed{<}$ $\frac{11}{15}$.

33. We compare $\frac{12}{27}$ to $\frac{4}{9}$ by converting $\frac{4}{9}$ to an equivalent fraction with denominator 27.

$$\frac{4}{9} \xrightarrow{\times 3} \frac{12}{27}$$

$\frac{12}{27} = \frac{12}{27}$, so $\frac{12}{27}$ $\boxed{=}$ $\frac{4}{9}$.

— *or* —

We compare $\frac{12}{27}$ to $\frac{4}{9}$ by simplifying $\frac{12}{27}$.

$$\frac{12}{27} \xrightarrow{\div 3} \frac{4}{9}$$

$\frac{4}{9} = \frac{4}{9}$, so $\frac{12}{27}$ $\boxed{=}$ $\frac{4}{9}$.

34. We compare $\frac{15}{23}$ to $\frac{5}{8}$ by converting $\frac{5}{8}$ to an equivalent fraction with numerator 15.

$$\frac{5}{8} \xrightarrow{\times 3} \frac{15}{24}$$

Twenty-thirds are larger than twenty-fourths, so $\frac{15}{23} > \frac{15}{24}$. Therefore, $\frac{15}{23} \boxed{>} \frac{5}{8}$.

35. We compare $\frac{40}{55}$ to $\frac{7}{11}$ by converting $\frac{7}{11}$ to an equivalent fraction with denominator 55.

$$\frac{7}{11} \xrightarrow{\times 5} \frac{35}{55}$$

$\frac{40}{55} > \frac{35}{55}$, so $\frac{40}{55} \boxed{>} \frac{7}{11}$.

— or —

We compare $\frac{40}{55}$ to $\frac{7}{11}$ by simplifying $\frac{40}{55}$.

$$\frac{40}{55} \xrightarrow{\div 5} \frac{8}{11}$$

$\frac{8}{11} > \frac{7}{11}$, so $\frac{40}{55} \boxed{>} \frac{7}{11}$.

36. We compare $\frac{2}{3}$ to $\frac{32}{45}$ by converting $\frac{2}{3}$ to an equivalent fraction with denominator 45.

$$\frac{2}{3} \xrightarrow{\times 15} \frac{30}{45}$$

$\frac{30}{45} < \frac{32}{45}$, so $\frac{2}{3} \boxed{<} \frac{32}{45}$.

— or —

We compare $\frac{2}{3}$ to $\frac{32}{45}$ by converting $\frac{2}{3}$ to an equivalent fraction with numerator 32.

$$\frac{2}{3} \xrightarrow{\times 16} \frac{32}{48}$$

Forty-eighths are smaller than forty-fifths, so $\frac{32}{48} < \frac{32}{45}$. Therefore, $\frac{2}{3} \boxed{<} \frac{32}{45}$.

37. We compare $\frac{45}{81}$ to $\frac{5}{7}$ by converting $\frac{5}{7}$ to an equivalent fraction with numerator 45.

$$\frac{5}{7} \xrightarrow{\times 9} \frac{45}{63}$$

Eighty-firsts are smaller than sixty-thirds, so $\frac{45}{81} < \frac{45}{63}$. Therefore, $\frac{45}{81} \boxed{<} \frac{5}{7}$.

— or —

We compare $\frac{45}{81}$ to $\frac{5}{7}$ by simplifying $\frac{45}{81}$.

$$\frac{45}{81} \xrightarrow{\div 9} \frac{5}{9}$$

Ninths are smaller than sevenths, so $\frac{5}{9} < \frac{5}{7}$. Therefore, $\frac{45}{81} \boxed{<} \frac{5}{7}$.

38. We compare $\frac{25}{39}$ to $\frac{8}{13}$ by converting $\frac{8}{13}$ to an equivalent fraction with denominator 39.

$$\frac{8}{13} \xrightarrow{\times 3} \frac{24}{39}$$

$\frac{25}{39} > \frac{24}{39}$, so $\frac{25}{39} \boxed{>} \frac{8}{13}$.

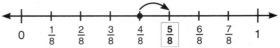

Addition   44-45

39. Adding 4 eighths to 1 eighth, we get a total of $4+1=5$ eighths. So, $\frac{4}{8}+\frac{1}{8}=\boldsymbol{\frac{5}{8}}$.

We can show this addition on the number line. Starting at $\frac{4}{8}$, we move right a distance of $\frac{1}{8}$ to $\frac{5}{8}$.

40. Adding 2 fifths to 2 fifths, we get a total of $2+2=4$ fifths. So, $\frac{2}{5}+\frac{2}{5}=\boldsymbol{\frac{4}{5}}$.

41. Adding 2 ninths to 5 ninths, we get a total of $2+5=7$ ninths. So, $\frac{2}{9}+\frac{5}{9}=\boldsymbol{\frac{7}{9}}$.

42. Adding 1 sixth to 5 sixths, we get a total of $1+5=6$ sixths. So, $\frac{1}{6}+\frac{5}{6}=\frac{6}{6}$. Since $\frac{6}{6}$ equals 1, we have $\frac{1}{6}+\frac{5}{6}=\boldsymbol{1}$.

43. Adding 4 elevenths to 3 elevenths, we get a total of $4+3=7$ elevenths. So, $\frac{4}{11}+\frac{3}{11}=\boldsymbol{\frac{7}{11}}$.

44. Adding 3 tenths to 6 tenths, we get a total of $3+6=9$ tenths. So, $\frac{3}{10}+\frac{6}{10}=\boldsymbol{\frac{9}{10}}$.

45. $\frac{4}{6}+\frac{1}{6}=\boldsymbol{\frac{5}{6}}$.

46. $\frac{1}{11}+\frac{1}{11}=\boldsymbol{\frac{2}{11}}$.

47. $\frac{2}{5}+\frac{1}{5}=\boldsymbol{\frac{3}{5}}$.

48. $\frac{4}{15}+\frac{8}{15}=\frac{12}{15}$. We simplify to get $\frac{12}{15}=\boldsymbol{\frac{4}{5}}$.

49. $\frac{2}{7}+\frac{3}{7}=\boldsymbol{\frac{5}{7}}$.

50. $\frac{7}{12}+\frac{1}{12}=\frac{8}{12}$. We simplify to get $\frac{8}{12}=\boldsymbol{\frac{2}{3}}$.

51. $\frac{3}{8}+\frac{5}{8}=\frac{8}{8}$. We simplify to get $\frac{8}{8}=\boldsymbol{1}$.

52. $\frac{2}{9}+\frac{4}{9}=\frac{6}{9}$. We simplify to get $\frac{6}{9}=\boldsymbol{\frac{2}{3}}$.

53. On April 5th, $\frac{1}{10}+\frac{7}{10}=\frac{8}{10}$ inches of rain fell. We simplify to get $\frac{8}{10}=\boldsymbol{\frac{4}{5}}$ inches.

54. Ally mixes the granola, chocolate bits, and mixed nuts to make $\frac{3}{8}+\frac{1}{8}+\frac{3}{8}=\boldsymbol{\frac{7}{8}}$ cups of trail mix.

Mixed Numbers   46-49

55. To find out the size of each serving, we divide the total number of ounces by the number of servings.

$15 \div 6$ has quotient 2 and remainder 3.

Each serving is at least 2 ounces. However, there are 3 remaining ounces to divide among the 6 servings. So, each serving receives an additional $3 \div 6 = \frac{3}{6}$ ounces.

All together, there are $2+\frac{3}{6}=2\frac{3}{6}=\mathbf{2\frac{1}{2}}$ **ounces** of chips in each serving.

We write $15\div 6 = \frac{15}{6}=\frac{12}{6}+\frac{3}{6}=2+\frac{3}{6}=2\frac{3}{6}=2\frac{1}{2}$.

56. A square has four equal sides. To find the length of each side, we divide the perimeter by the number of sides.

$39\div 4$ has quotient 9 and remainder 3.

Each side of the square is at least 9 inches long. However, there are 3 remaining inches to divide among the 4 sides. So, each side is an additional $3\div 4 = \frac{3}{4}$ inches long.

All together, the side length of the square is $9+\frac{3}{4}=\mathbf{9\frac{3}{4}}$ **inches.**

We write $39\div 4 = \frac{39}{4}=\frac{36}{4}+\frac{3}{4}=9+\frac{3}{4}=9\frac{3}{4}$.

57. To find the length of each piece of rope, we divide the total length of the rope by the number of pieces cut.

$22\div 3$ has quotient 7 and remainder 1.

Each of the three pieces is at least 7 inches long. However, there is 1 remaining inch of the original rope to divide among the 3 pieces. So, each piece is an additional $1\div 3 = \frac{1}{3}$ inches long.

All together, each piece of rope is $7+\frac{1}{3}=\mathbf{7\frac{1}{3}}$ **inches** long.

We write $22\div 3 = \frac{22}{3}=\frac{21}{3}+\frac{1}{3}=7+\frac{1}{3}=7\frac{1}{3}$.

58. To find the weight of each bag of cereal, we divide the total number of ounces by the number of bags.

$21\div 4$ has quotient 5 and remainder 1.

Each bag weighs at least 5 ounces. However, there is 1 remaining ounce to divide among the 4 bags. So, each bag receives an additional $1\div 4 = \frac{1}{4}$ ounces.

All together, each bag weighs $5+\frac{1}{4}=\mathbf{5\frac{1}{4}}$ **ounces.**

We write $21\div 4 = \frac{21}{4}=\frac{20}{4}+\frac{1}{4}=5+\frac{1}{4}=5\frac{1}{4}$.

59. Since 14 is half of 28, we know Brenda uses 14 ounces of wax to make the 2 large candles. Then, she has $28-14=14$ ounces left to make the small candles.

To find the number of ounces used for each small candle, we divide the total number of ounces used to make the small candles by the number of small candles.

$14\div 4$ has quotient 3 and remainder 2.

So, each small candle uses at least 3 ounces of wax. However, there are 2 remaining ounces of wax to divide among the 4 small candles. So, each small candle receives an additional $2\div 4 = \frac{2}{4}$ ounces of wax.

All together, there are $3+\frac{2}{4}=3\frac{2}{4}=\mathbf{3\frac{1}{2}}$ **ounces** of wax in each small candle.

We write $14\div 4 = \frac{14}{4}=\frac{12}{4}+\frac{2}{4}=3+\frac{2}{4}=3\frac{2}{4}=3\frac{1}{2}$.

60. Since $5=\frac{50}{10}$, we know $\frac{53}{10}$ is three tenths more than 5.
So, $\frac{53}{10}=\frac{50}{10}+\frac{3}{10}=5+\frac{3}{10}=\mathbf{5\frac{3}{10}}$.

61. Since $7=\frac{63}{9}$, we know $\frac{70}{9}$ is seven ninths more than 7.
So, $\frac{70}{9}=\frac{63}{9}+\frac{7}{9}=7+\frac{7}{9}=\mathbf{7\frac{7}{9}}$.

62. Since $8=\frac{32}{4}$, we know $\frac{35}{4}$ is three fourths more than 8.
So, $\frac{35}{4}=\frac{32}{4}+\frac{3}{4}=8+\frac{3}{4}=\mathbf{8\frac{3}{4}}$.

63. Since $4=\frac{32}{8}$, we know $\frac{34}{8}$ is two eighths more than 4.
So, $\frac{34}{8}=\frac{32}{8}+\frac{2}{8}=4+\frac{2}{8}=4\frac{2}{8}=\mathbf{4\frac{1}{4}}$.

— or —

We first simplify $\frac{34}{8}=\frac{17}{4}$.

Then, since $4=\frac{16}{4}$, we know $\frac{17}{4}$ is one fourth more than 4.
So, $\frac{34}{8}=\frac{17}{4}=\frac{16}{4}+\frac{1}{4}=4+\frac{1}{4}=\mathbf{4\frac{1}{4}}$.

64. Since $15=\frac{90}{6}$, we know $\frac{92}{6}$ is two sixths more than 15.
So, $\frac{92}{6}=\frac{90}{6}+\frac{2}{6}=15+\frac{2}{6}=15\frac{2}{6}=\mathbf{15\frac{1}{3}}$.

— or —

We first simplify $\frac{92}{6}=\frac{46}{3}$.

Then, since $15=\frac{45}{3}$, we know $\frac{46}{3}$ is one third more than 15. So, $\frac{46}{3}=\frac{45}{3}+\frac{1}{3}=15+\frac{1}{3}=\mathbf{15\frac{1}{3}}$.

65. Since $5=\frac{60}{12}$, we know $\frac{68}{12}$ is eight twelfths more than 5.
So, $\frac{68}{12}=\frac{60}{12}+\frac{8}{12}=5+\frac{8}{12}=5\frac{8}{12}=\mathbf{5\frac{2}{3}}$.

— or —

We first simplify $\frac{68}{12}=\frac{17}{3}$.

Then, since $5=\frac{15}{3}$, we know $\frac{17}{3}$ is two thirds more than 5.
So, $\frac{17}{3}=\frac{15}{3}+\frac{2}{3}=5+\frac{2}{3}=\mathbf{5\frac{2}{3}}$.

66. Since $27=\frac{135}{5}$, we know $\frac{137}{5}$ is two fifths more than 27.
So, $\frac{137}{5}=\frac{135}{5}+\frac{2}{5}=27+\frac{2}{5}=\mathbf{27\frac{2}{5}}$.

67. Since $8=\frac{112}{14}$, we know $\frac{119}{14}$ is seven fourteenths more than 8. So, $\frac{119}{14}=\frac{112}{14}+\frac{7}{14}=8+\frac{7}{14}=8\frac{7}{14}=\mathbf{8\frac{1}{2}}$.

— or —

We first simplify $\frac{119}{14}=\frac{17}{2}$.

Then, since $8=\frac{16}{2}$, we know $\frac{17}{2}$ is one half more than 8.
So, $\frac{17}{2}=\frac{16}{2}+\frac{1}{2}=8+\frac{1}{2}=\mathbf{8\frac{1}{2}}$.

68. We know that $\frac{39}{7}=5\frac{4}{7}$ is between 5 and 6.

We also know that $\frac{44}{3}=14\frac{2}{3}$ is between 14 and 15.

So, the whole numbers between $\frac{39}{7}$ and $\frac{44}{3}$ are 6, 7, 8, 9, 10, 11, 12, 13, and 14.

All together, we count **9** whole numbers.

69. We write each of the given fractions as a mixed number.

$\frac{79}{5}=15\frac{4}{5}$ $\frac{17}{11}=1\frac{6}{11}$ $\frac{111}{8}=13\frac{7}{8}$ $\frac{51}{4}=12\frac{3}{4}$

Three numbers are greater than 10, and $12\frac{3}{4}$ is the smallest of the three numbers. So, of the three numbers greater than 10, we know $\mathbf{12\frac{3}{4}}$ is the closest to 10.

We compare the distance from $1\frac{6}{11}$ to 10 and the distance from 10 to $12\frac{3}{4}$.

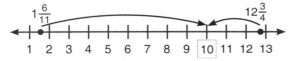

$12\frac{3}{4}$ is less than 3 units from 10, while $1\frac{6}{11}$ is more than 8 units from 10. So, among these fractions, $12\frac{3}{4} = \frac{51}{4}$ is closest to 10. $\frac{79}{5}$ $\frac{17}{11}$ $\frac{111}{8}$ $\boxed{\frac{51}{4}}$

70. To make these numbers easier to compare, we write each of the given fractions as a mixed number.

$\frac{59}{9} = 6\frac{5}{9}$ $\frac{27}{5} = 5\frac{2}{5}$ $\frac{45}{11} = 4\frac{1}{11}$ $\frac{31}{8} = 3\frac{7}{8}$

Now, the whole-number parts of these mixed numbers make them easy to order: $3\frac{7}{8} < 4\frac{1}{11} < 5\frac{2}{5} < 6\frac{5}{9}$.

We write the fractions in order from least to greatest: $\frac{31}{8}, \frac{45}{11}, \frac{27}{5},$ and $\frac{59}{9}$.

71. We convert both numbers in the sum into mixed numbers: $\frac{50}{6} = 8\frac{2}{6} = 8\frac{1}{3}$ and $\frac{65}{7} = 9\frac{2}{7}$.

So, $\frac{50}{6} + \frac{65}{7} = 8\frac{1}{3} + 9\frac{2}{7}$.

Since $8\frac{1}{3} = 8 + \frac{1}{3}$ and $9\frac{2}{7} = 9 + \frac{2}{7}$, we rewrite the sum $8\frac{1}{3} + 9\frac{2}{7}$ as $8 + \frac{1}{3} + 9 + \frac{2}{7}$.

Then, rearranging and grouping the terms in the sum, we get $8 + \frac{1}{3} + 9 + \frac{2}{7} = (8+9) + \left(\frac{1}{3} + \frac{2}{7}\right) = 17 + \left(\frac{1}{3} + \frac{2}{7}\right)$.

Since $\frac{1}{3}$ and $\frac{2}{7}$ are each less than $\frac{1}{2}$, their sum is less than 1. So, $17 + \left(\frac{1}{3} + \frac{2}{7}\right)$ is less than 18.

Therefore, $\frac{50}{6} + \frac{65}{7}$ is between **17 and 18**.

72. We look at the sum on the left first. We convert the fractions into mixed numbers: $\frac{33}{4} + \frac{23}{7} = 8\frac{1}{4} + 3\frac{2}{7}$.
Next, we rearrange and group the terms to get
$8 + \frac{1}{4} + 3 + \frac{2}{7} = (8+3) + \left(\frac{1}{4} + \frac{2}{7}\right) = 11 + \left(\frac{1}{4} + \frac{2}{7}\right)$.
So, this sum is greater than 11.

Next, we look at the sum on the right. We convert the fractions to mixed numbers: $\frac{15}{2} + \frac{33}{16} = 7\frac{1}{2} + 2\frac{1}{16}$.

Since $7\frac{1}{2}$ is less than 8 and $2\frac{1}{16}$ is less than 3, we know $7\frac{1}{2} + 2\frac{1}{16}$ is less than $8 + 3 = 11$.

So, $\frac{33}{4} + \frac{23}{7}$ is greater than 11, and $\frac{15}{2} + \frac{33}{16}$ is less than 11. Therefore, $\frac{33}{4} + \frac{23}{7}$ must be the greater sum, and we have $\frac{33}{4} + \frac{23}{7} \boxed{>} \frac{15}{2} + \frac{33}{16}$.

73. The fractional part of the mixed number has denominator 2. So, we write a fraction equal to 5 that has the same denominator: $5 = \frac{10}{2}$.
$5\frac{1}{2}$ is one half more than 5, so $5\frac{1}{2} = 5 + \frac{1}{2} = \frac{10}{2} + \frac{1}{2} = \frac{11}{2}$.

74. $7\frac{2}{3}$ is two thirds more than 7, and $7 = \frac{21}{3}$.
So, $7\frac{2}{3} = 7 + \frac{2}{3} = \frac{21}{3} + \frac{2}{3} = \frac{23}{3}$.

75. $11\frac{6}{7}$ is six sevenths more than 11, and $11 = \frac{77}{7}$.
So, $11\frac{6}{7} = 11 + \frac{6}{7} = \frac{77}{7} + \frac{6}{7} = \frac{83}{7}$.

76. $4\frac{4}{10}$ is four tenths more than $4 = \frac{40}{10}$.
So, $4\frac{4}{10} = 4 + \frac{4}{10} = \frac{40}{10} + \frac{4}{10} = \frac{44}{10} = \frac{22}{5}$.

— or —

We begin by simplifying the fractional part of $4\frac{4}{10} = 4\frac{2}{5}$.
$4\frac{2}{5}$ is two fifths more than $4 = \frac{20}{5}$.
So, $4\frac{4}{10} = 4\frac{2}{5} = 4 + \frac{2}{5} = \frac{20}{5} + \frac{2}{5} = \frac{22}{5}$.

77. $6\frac{4}{5}$ is four fifths more than 6, and $6 = \frac{30}{5}$.
So, $6\frac{4}{5} = 6 + \frac{4}{5} = \frac{30}{5} + \frac{4}{5} = \frac{34}{5}$.

78. $9\frac{4}{9}$ is four ninths more than 9, and $9 = \frac{81}{9}$.
So, $9\frac{4}{9} = 9 + \frac{4}{9} = \frac{81}{9} + \frac{4}{9} = \frac{85}{9}$.

FRACTIONS
Skip-Counting   50-51

79. To get from $\frac{1}{11}$ to $\frac{2}{11}$, we add $\frac{1}{11}$. Similarly, to get from $\frac{2}{11}$ to $\frac{3}{11}$, we add $\frac{1}{11}$. So, we add $\frac{1}{11}$ to each number to get the next number.

$\frac{1}{11}, \frac{2}{11}, \frac{3}{11}, \boxed{\frac{4}{11}}, \boxed{\frac{5}{11}}, \boxed{\frac{6}{11}}, \boxed{\frac{7}{11}}$

80. We add $\frac{1}{8}$ to each number to get the next number.

$\frac{1}{8}, \frac{2}{8}, \frac{3}{8}, \boxed{\frac{4}{8}}, \boxed{\frac{5}{8}}, \boxed{\frac{6}{8}}, \boxed{\frac{7}{8}}$

Then, we rewrite the pattern, simplifying when possible.

$\frac{1}{8}, \frac{1}{4}, \frac{3}{8}, \frac{1}{2}, \frac{5}{8}, \frac{3}{4}, \frac{7}{8}$

81. We add $\frac{1}{9}$ to each number to get the next number.

$\frac{4}{9}, \frac{5}{9}, \frac{6}{9}, \boxed{\frac{7}{9}}, \boxed{\frac{8}{9}}, \boxed{\frac{9}{9}}, \boxed{\frac{10}{9}}$

Then, we rewrite the pattern, simplifying and using whole or mixed numbers when possible.

$\frac{4}{9}, \frac{5}{9}, \frac{2}{3}, \frac{7}{9}, \frac{8}{9}, 1, 1\frac{1}{9}$

82. We add $\frac{1}{4}$ to each number to get the next number.

$\frac{3}{4}, \frac{4}{4}, \frac{5}{4}, \boxed{\frac{6}{4}}, \boxed{\frac{7}{4}}, \boxed{\frac{8}{4}}, \boxed{\frac{9}{4}}$

Then, we rewrite the pattern, simplifying and using whole or mixed numbers when possible.

$\frac{3}{4}, 1, 1\frac{1}{4}, 1\frac{1}{2}, 1\frac{3}{4}, 2, 2\frac{1}{4}$

83. We add $\frac{2}{12}$ to each number to get the next number.

$\frac{3}{12}, \frac{5}{12}, \frac{7}{12}, \boxed{\frac{9}{12}}, \boxed{\frac{11}{12}}, \boxed{\frac{13}{12}}, \boxed{\frac{15}{12}}$

Then, we rewrite the pattern, simplifying and using whole or mixed numbers when possible.

$\frac{1}{4}, \frac{5}{12}, \frac{7}{12}, \frac{3}{4}, \frac{11}{12}, 1\frac{1}{12}, 1\frac{1}{4}$

84. We add $\frac{1}{35}$ to each number to get the next number.

$$\frac{2}{35}, \frac{3}{35}, \frac{4}{35}, \frac{5}{35}, \frac{6}{35}, \frac{7}{35}, \frac{8}{35}$$

Then, we rewrite the pattern, simplifying when possible.

$$\frac{2}{35}, \frac{3}{35}, \frac{4}{35}, \boxed{\frac{1}{7}}, \boxed{\frac{6}{35}}, \boxed{\frac{1}{5}}, \boxed{\frac{8}{35}}$$

85. In order to find the pattern, we convert the existing fractions into fractions with the same denominator.

$\frac{1}{3} = \frac{2}{6}$, so we rewrite the pattern as

$$\frac{1}{6}, \frac{2}{6}, \boxed{}, \boxed{}, \frac{5}{6}, \boxed{}, \boxed{}.$$

To get from $\frac{1}{6}$ to $\frac{2}{6}$, we add $\frac{1}{6}$. Then, to get from $\frac{2}{6}$ to $\frac{5}{6}$, we can add $\frac{1}{6}$ three times.

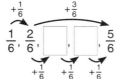

So, to complete the skip-counting pattern, we add $\frac{1}{6}$ to each number to get the next number.

$$\frac{1}{6}, \frac{2}{6}, \frac{3}{6}, \frac{4}{6}, \frac{5}{6}, \frac{6}{6}, \frac{7}{6}$$

Then, we rewrite the pattern, simplifying and using whole or mixed numbers when possible.

$$\frac{1}{6}, \frac{1}{3}, \boxed{\frac{1}{2}}, \boxed{\frac{2}{3}}, \frac{5}{6}, \boxed{1}, \boxed{1\frac{1}{6}}$$

86. In order to find the pattern, we convert the existing fractions into fractions with the same denominator.

Both $\frac{1}{3}$ and $\frac{3}{5}$ can be written with denominator 15: $\frac{1}{3} = \frac{5}{15}$ and $\frac{3}{5} = \frac{9}{15}$. So, we rewrite the pattern as

$$\frac{5}{15}, \frac{9}{15}, \frac{13}{15}, \boxed{}, \boxed{}, \boxed{}.$$

To get from $\frac{5}{15}$ to $\frac{9}{15}$, we add $\frac{4}{15}$. Similarly, to get from $\frac{9}{15}$ to $\frac{13}{15}$, we add $\frac{4}{15}$. So, we add $\frac{4}{15}$ to each number to get the next number.

$$\frac{5}{15}, \frac{9}{15}, \frac{13}{15}, \frac{17}{15}, \frac{21}{15}, \frac{25}{15}, \frac{29}{15}$$

Then, we rewrite the pattern, simplifying when possible.

$$\frac{1}{3}, \frac{3}{5}, \frac{13}{15}, \boxed{1\frac{2}{15}}, \boxed{1\frac{2}{5}}, \boxed{1\frac{2}{3}}, \boxed{1\frac{14}{15}}$$

FRACTIONS
Maze Escape 52-53

87. Our escape path begins at $\frac{1}{20}$, and we continue adding $\frac{2}{20}$ to escape the maze:

$$\frac{1}{20}, \frac{3}{20}, \frac{5}{20}, \frac{7}{20}, \frac{9}{20}, \frac{11}{20}, \frac{13}{20}, \frac{15}{20}, \frac{17}{20}, \frac{19}{20}.$$

88. We continue adding $\frac{6}{11}$ to escape the maze.

89. We continue adding $\frac{6}{13}$ to escape the maze.

90. We continue adding $\frac{3}{19}$ to escape the maze.

91. We continue adding $\frac{2}{3}$ to escape the maze.

92. We continue adding $\frac{3}{7}$ to escape the maze.

93. To begin, we write all of the numbers as fractions with the same denominator. We can write all of the numbers in the grid as fractions with denominator 9.

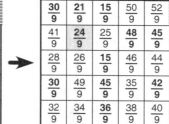

Our escape path begins at $\frac{24}{9}$, and we continue adding $\frac{2}{9}$ to escape the maze.

$\frac{10}{3}$	$\frac{7}{3}$	$\frac{5}{3}$	$\frac{50}{9}$	$\frac{52}{9}$
$\frac{41}{9}$	$\frac{8}{3}$	$\frac{25}{9}$	$\frac{16}{3}$	5
$\frac{28}{9}$	$\frac{26}{9}$	$\frac{5}{3}$	$\frac{46}{9}$	$\frac{44}{9}$
$\frac{10}{3}$	$\frac{49}{9}$	$\frac{15}{3}$	$\frac{35}{9}$	$\frac{14}{3}$
$\frac{32}{9}$	$\frac{34}{9}$	4	$\frac{38}{9}$	$\frac{40}{9}$

94. To begin, we write all of the numbers as fractions with the same denominator. We can write all of the numbers in the grid as fractions with denominator 4.

$\frac{28}{2}$	15	$\frac{63}{4}$	$\frac{33}{2}$	$\frac{31}{2}$
$\frac{25}{2}$	$\frac{57}{4}$	10	$\frac{69}{4}$	$\frac{21}{2}$
$\frac{55}{4}$	$\frac{27}{2}$	$\frac{39}{4}$	18	$\frac{81}{4}$
$\frac{47}{4}$	$\frac{51}{4}$	$\frac{21}{2}$	$\frac{75}{4}$	$\frac{39}{2}$
$\frac{26}{2}$	12	$\frac{45}{4}$	$\frac{83}{4}$	$\frac{35}{2}$

→

$\frac{56}{4}$	$\frac{60}{4}$	$\frac{63}{4}$	$\frac{66}{4}$	$\frac{62}{4}$
$\frac{50}{4}$	$\frac{57}{4}$	$\frac{40}{4}$	$\frac{69}{4}$	$\frac{42}{4}$
$\frac{55}{4}$	$\frac{54}{4}$	$\frac{39}{4}$	$\frac{72}{4}$	$\frac{81}{4}$
$\frac{47}{4}$	$\frac{51}{4}$	$\frac{42}{4}$	$\frac{75}{4}$	$\frac{78}{4}$
$\frac{52}{4}$	$\frac{48}{4}$	$\frac{45}{4}$	$\frac{83}{4}$	$\frac{70}{4}$

Our escape path begins at $\frac{39}{4}$, and we continue adding $\frac{3}{4}$ to escape the maze.

$\frac{28}{2}$	15	$\frac{63}{4}$	$\frac{33}{2}$	$\frac{31}{2}$
$\frac{25}{2}$	$\frac{57}{4}$	10	$\frac{69}{4}$	$\frac{21}{2}$
$\frac{55}{4}$	$\frac{27}{2}$	$\frac{39}{4}$	18	$\frac{81}{4}$
$\frac{47}{4}$	$\frac{51}{4}$	$\frac{21}{2}$	$\frac{75}{4}$	$\frac{39}{2}$
$\frac{26}{2}$	12	$\frac{45}{4}$	$\frac{83}{4}$	$\frac{35}{2}$

Adding Mixed Numbers 54-55

95. We add the whole numbers and fractions separately:

$6\frac{5}{11} + 4\frac{3}{11} = \left(6+\frac{5}{11}\right) + \left(4+\frac{3}{11}\right)$
$= (6+4) + \left(\frac{5}{11}+\frac{3}{11}\right)$
$= 10 + \frac{8}{11}$
$= \mathbf{10\frac{8}{11}}.$

— *or* —

We stack the numbers and add.

$\begin{array}{r} 6\frac{5}{11} \\ +4\frac{3}{11} \\ \hline 10\frac{8}{11} \end{array}$

In the next problems, we display the addition horizontally. You may have instead stacked the numbers to organize your work.

96. $5\frac{2}{7} + 6\frac{3}{7} = \left(5+\frac{2}{7}\right) + \left(6+\frac{3}{7}\right)$
$= (5+6) + \left(\frac{2}{7}+\frac{3}{7}\right)$
$= 11 + \frac{5}{7}$
$= \mathbf{11\frac{5}{7}}.$

97. $9\frac{4}{13} + 3\frac{7}{13} = \left(9+\frac{4}{13}\right) + \left(3+\frac{7}{13}\right)$
$= (9+3) + \left(\frac{4}{13}+\frac{7}{13}\right)$
$= 12 + \frac{11}{13}$
$= \mathbf{12\frac{11}{13}}.$

98. $2\frac{2}{5} + 1\frac{1}{5} = \left(2+\frac{2}{5}\right) + \left(1+\frac{1}{5}\right)$
$= (2+1) + \left(\frac{2}{5}+\frac{1}{5}\right)$
$= 3 + \frac{3}{5}$
$= \mathbf{3\frac{3}{5}}.$

99. $8\frac{6}{11} + 4\frac{9}{11} = \left(8+\frac{6}{11}\right) + \left(4+\frac{9}{11}\right)$
$= (8+4) + \left(\frac{6}{11}+\frac{9}{11}\right)$
$= 12 + \frac{15}{11}.$

Since $\frac{15}{11}$ is not less than 1, we regroup. We write $\frac{15}{11}$ as $\frac{11}{11} + \frac{4}{11} = 1 + \frac{4}{11}$, and continue adding.

$8\frac{6}{11} + 4\frac{9}{11} = 12 + \frac{15}{11}$
$= 12 + \frac{11}{11} + \frac{4}{11}$
$= 12 + 1 + \frac{4}{11}$
$= 13 + \frac{4}{11}$
$= \mathbf{13\frac{4}{11}}.$

100. $2\frac{5}{9} + 8\frac{4}{9} = \left(2+\frac{5}{9}\right) + \left(8+\frac{4}{9}\right)$
$= (2+8) + \left(\frac{5}{9}+\frac{4}{9}\right)$
$= 10 + \frac{9}{9}$
$= 10 + 1$
$= \mathbf{11}.$

101. $3\frac{4}{5} + 2\frac{2}{5} = \left(3+\frac{4}{5}\right) + \left(2+\frac{2}{5}\right)$
$= (3+2) + \left(\frac{4}{5}+\frac{2}{5}\right)$
$= 5 + \frac{6}{5}$
$= 5 + \frac{5}{5} + \frac{1}{5}$
$= 5 + 1 + \frac{1}{5}$
$= 6 + \frac{1}{5}$
$= \mathbf{6\frac{1}{5}}.$

102. $3\frac{2}{5} + 9\frac{1}{5} = \left(3+\frac{2}{5}\right) + \left(9+\frac{1}{5}\right)$
$= (3+9) + \left(\frac{2}{5}+\frac{1}{5}\right)$
$= 12 + \frac{3}{5}$
$= \mathbf{12\frac{3}{5}}.$

103. $8\frac{5}{6} + 3\frac{5}{6} = \left(8 + \frac{5}{6}\right) + \left(3 + \frac{5}{6}\right)$
$= (8+3) + \left(\frac{5}{6} + \frac{5}{6}\right)$
$= 11 + \frac{10}{6}$
$= 11 + \frac{6}{6} + \frac{4}{6}$
$= 11 + 1 + \frac{4}{6}$
$= 12 + \frac{4}{6}$
$= 12\frac{4}{6}$
$= \mathbf{12\frac{2}{3}}.$

104. $4\frac{10}{19} + 25\frac{15}{19} = \left(4 + \frac{10}{19}\right) + \left(25 + \frac{15}{19}\right)$
$= (4+25) + \left(\frac{10}{19} + \frac{15}{19}\right)$
$= 29 + \frac{25}{19}$
$= 29 + \frac{19}{19} + \frac{6}{19}$
$= 29 + 1 + \frac{6}{19}$
$= 30 + \frac{6}{19}$
$= \mathbf{30\frac{6}{19}}.$

105. $17\frac{7}{8} + 2\frac{7}{8} + 2\frac{7}{8} = \left(17 + \frac{7}{8}\right) + \left(2 + \frac{7}{8}\right) + \left(2 + \frac{7}{8}\right)$
$= (17+2+2) + \left(\frac{7}{8} + \frac{7}{8} + \frac{7}{8}\right)$
$= 21 + \frac{21}{8}$
$= 21 + \frac{16}{8} + \frac{5}{8}$
$= 21 + 2 + \frac{5}{8}$
$= 23 + \frac{5}{8}$
$= \mathbf{23\frac{5}{8}}.$

106. $6\frac{10}{13} + 22\frac{9}{13} + 8\frac{7}{13} = \left(6 + \frac{10}{13}\right) + \left(22 + \frac{9}{13}\right) + \left(8 + \frac{7}{13}\right)$
$= (6+22+8) + \left(\frac{10}{13} + \frac{9}{13} + \frac{7}{13}\right)$
$= 36 + \frac{26}{13}$
$= 36 + \frac{26}{13}$
$= 36 + 2$
$= \mathbf{38}.$

107. All together, Mr. Jones uses
$1\frac{2}{3} + 2\frac{2}{3} = (1+2) + \left(\frac{2}{3} + \frac{2}{3}\right)$
$= 3 + \frac{4}{3}$
$= 3 + \frac{3}{3} + \frac{1}{3}$
$= 3 + 1 + \frac{1}{3}$
$= \mathbf{4\frac{1}{3}}$ gallons of paint.

108. The perimeter of the triangle is
$2\frac{3}{4} + 2\frac{3}{4} + 2\frac{3}{4} = (2+2+2) + \left(\frac{3}{4} + \frac{3}{4} + \frac{3}{4}\right)$
$= 6 + \frac{9}{4}$
$= 6 + \frac{8}{4} + \frac{1}{4}$
$= 6 + 2 + \frac{1}{4}$
$= \mathbf{8\frac{1}{4}}$ inches.

109. The perimeter of the rectangle is
$5\frac{6}{7} + 2\frac{4}{7} + 5\frac{6}{7} + 2\frac{4}{7} = (5+2+5+2) + \left(\frac{6}{7} + \frac{4}{7} + \frac{6}{7} + \frac{4}{7}\right)$
$= 14 + \frac{20}{7}$
$= 14 + \frac{14}{7} + \frac{6}{7}$
$= 14 + 2 + \frac{6}{7}$
$= \mathbf{16\frac{6}{7}}$ inches.

Subtracting Fractions   56-57

110. Subtracting 1 eighth from 4 eighths leaves $4-1 = 3$ eighths. So, $\frac{4}{8} - \frac{1}{8} = \mathbf{\frac{3}{8}}$.

We can show this subtraction on the number line. Starting at $\frac{4}{8}$, we move left a distance of one eighth to $\frac{3}{8}$.

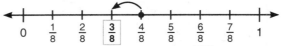

We check our answer with addition: $\boxed{\frac{3}{8}} + \frac{1}{8} = \frac{4}{8}$. ✓

111. Subtracting two fifths from three fifths leaves $3-2 = 1$ fifth. So, $\frac{3}{5} - \frac{2}{5} = \mathbf{\frac{1}{5}}$.

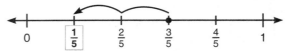

112. Subtracting 7 ninths from 8 ninths leaves $8-7 = 1$ ninth. So, $\frac{8}{9} - \frac{7}{9} = \mathbf{\frac{1}{9}}$.

113. Subtracting 4 sixths from 5 sixths leaves $5-4 = 1$ sixth. So, $\frac{5}{6} - \frac{4}{6} = \mathbf{\frac{1}{6}}$.

114. Subtracting 6 elevenths from 10 elevenths leaves $10-6 = 4$ elevenths. So, $\frac{10}{11} - \frac{6}{11} = \mathbf{\frac{4}{11}}$.

115. Subtracting 5 tenths from 8 tenths leaves $8-5 = 3$ tenths. So, $\frac{8}{10} - \frac{5}{10} = \mathbf{\frac{3}{10}}$.

116. $\frac{9}{13} - \frac{1}{13} = \mathbf{\frac{8}{13}}$.

117. $\frac{8}{9} - \frac{1}{9} = \mathbf{\frac{7}{9}}$.

118. $\frac{7}{12} - \frac{4}{12} = \frac{3}{12}$, which simplifies to $\mathbf{\frac{1}{4}}$.

119. $\frac{7}{8} - \frac{5}{8} = \frac{2}{8}$, which simplifies to $\mathbf{\frac{1}{4}}$.

120. $\frac{22}{15} - \frac{13}{15} = \frac{9}{15}$, which simplifies to $\mathbf{\frac{3}{5}}$.

121. $\frac{20}{9} - \frac{14}{9} = \frac{6}{9}$, which simplifies to $\mathbf{\frac{2}{3}}$.

122. The Beast Island dollar coin is $\frac{7}{25} - \frac{2}{25} = \frac{5}{25}$ ounces heavier than the Beast Island dime.
We simplify to get $\frac{5}{25} = \mathbf{\frac{1}{5}}$ ounces.

123. The distance from B to C is equal to the distance from A to C minus the distance from A to B.
So, the distance from B to C is $\frac{33}{16} - \frac{25}{16} = \frac{8}{16}$ inches.
We simplify to get $\frac{8}{16} = \mathbf{\frac{1}{2}}$ inches.

124. Alex must swim $\frac{25}{10} - \frac{17}{10} = \frac{8}{10}$ more miles to meet his goal.
We simplify to get $\frac{8}{10} = \mathbf{\frac{4}{5}}$ miles.

FRACTIONS
Constellation Puzzles — 58-59

125. We have $\frac{23}{5} = 4\frac{3}{5} = 3\frac{8}{5}$ and $\frac{26}{5} = 5\frac{1}{5} = 4\frac{6}{5}$.

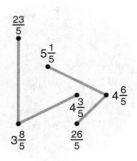

126. We have $\frac{11}{3} = 3\frac{2}{3} = 2\frac{5}{3}$ and $\frac{13}{3} = 4\frac{1}{3} = 3\frac{4}{3}$.

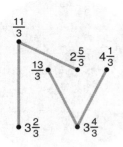

127. We have the following equivalent expressions:
$\frac{13}{5} = 2\frac{3}{5} = 1\frac{8}{5}$,
$\frac{16}{5} = 3\frac{1}{5} = 2\frac{6}{5}$, and
$\frac{14}{5} = 2\frac{4}{5} = 1\frac{9}{5}$.

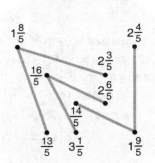

128. We have the following equivalent expressions:
$\frac{19}{8} = 2\frac{3}{8} = 1\frac{11}{8}$,
$\frac{21}{8} = 2\frac{5}{8} = 1\frac{13}{8}$, and
$\frac{23}{8} = 2\frac{7}{8} = 1\frac{15}{8}$.

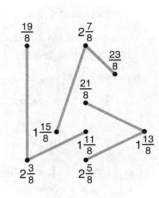

129. We have the following equivalent expressions:
$\frac{29}{9} = 3\frac{2}{9} = 2\frac{11}{9}$,
$\frac{35}{9} = 3\frac{8}{9} = 2\frac{17}{9}$, and
$\frac{32}{9} = 3\frac{5}{9} = 2\frac{14}{9}$.

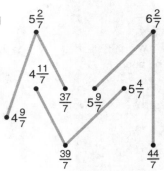

130. We have the following equivalent expressions:
$\frac{37}{7} = 5\frac{2}{7} = 4\frac{9}{7}$,
$\frac{39}{7} = 5\frac{4}{7} = 4\frac{11}{7}$, and
$\frac{44}{7} = 6\frac{2}{7} = 5\frac{9}{7}$.

FRACTIONS
Subtracting Mixed Numbers — 60-61

131. We first convert each mixed number to a fraction:
$8\frac{3}{7} = \frac{56}{7} + \frac{3}{7} = \frac{59}{7}$ and $2\frac{1}{7} = \frac{14}{7} + \frac{1}{7} = \frac{15}{7}$.

Then, we subtract: $8\frac{3}{7} - 2\frac{1}{7} = \frac{59}{7} - \frac{15}{7} = \frac{44}{7}$.

Converting our final answer to a mixed number, we get $\frac{44}{7} = \frac{42}{7} + \frac{2}{7} = 6\frac{2}{7}$. Therefore, $8\frac{3}{7} - 2\frac{1}{7} = \mathbf{6\frac{2}{7}}$.

— *or* —

We stack the mixed numbers and subtract.

$\begin{array}{r} 8\frac{3}{7} \\ -2\frac{1}{7} \\ \hline 6\frac{2}{7} \end{array}$ $8 - 2 = 6$, and $\frac{3}{7} - \frac{1}{7} = \frac{2}{7}$.
So, $8\frac{3}{7} - 2\frac{1}{7} = \mathbf{6\frac{2}{7}}$.

We check our answer with addition: $\boxed{6\frac{2}{7}} + 2\frac{1}{7} = 8\frac{3}{7}$. ✓

132. $\begin{array}{r} 9\frac{7}{11} \\ -5\frac{3}{11} \\ \hline 4\frac{4}{11} \end{array}$ $9 - 5 = 4$, and $\frac{7}{11} - \frac{3}{11} = \frac{4}{11}$.
So, $9\frac{7}{11} - 5\frac{3}{11} = \mathbf{4\frac{4}{11}}$.

133. $\begin{array}{r} 3\frac{5}{9} \\ -1\frac{5}{9} \\ \hline 2\frac{0}{9} \end{array}$ $3 - 1 = 2$, and $\frac{5}{9} - \frac{5}{9} = \frac{0}{9} = 0$.
So, $3\frac{5}{9} - 1\frac{5}{9} = 2\frac{0}{9} = \mathbf{2}$.

Beast Academy Practice 4C
Fractions Chapter 8 Solutions

134. $7\frac{9}{10}$
$-2\frac{3}{10}$
$\overline{5\frac{6}{10}}$

$7-2=5$, and $\frac{9}{10}-\frac{3}{10}=\frac{6}{10}$.
So, $7\frac{9}{10}-2\frac{3}{10}=5\frac{6}{10}$.
In simplest form, we have $5\frac{6}{10}=\mathbf{5\frac{3}{5}}$.

135. $5\frac{4}{5}$
$-2\frac{3}{5}$
$\overline{3\frac{1}{5}}$

$5-2=3$, and $\frac{4}{5}-\frac{3}{5}=\frac{1}{5}$.
So, $5\frac{4}{5}-2\frac{3}{5}=\mathbf{3\frac{1}{5}}$.

136. $6\frac{11}{15}$
$-4\frac{2}{15}$
$\overline{2\frac{9}{15}}$

$6-4=2$, and $\frac{11}{15}-\frac{2}{15}=\frac{9}{15}$.
So, $6\frac{11}{15}-4\frac{2}{15}=2\frac{9}{15}$.
In simplest form, we have $2\frac{9}{15}=\mathbf{2\frac{3}{5}}$.

137. $10\frac{4}{9}$
$-6\frac{5}{9}$

$10-6=4$, but we cannot take away 5 ninths from 4 ninths.

So, we regroup $10\frac{4}{9}$. We take $1=\frac{9}{9}$ from the 10 and add it to $\frac{4}{9}$:
$10\frac{4}{9}=9+\frac{9}{9}+\frac{4}{9}=9+\frac{13}{9}$.

So, $10\frac{4}{9}=9\frac{13}{9}$. Now, we subtract.

$9\frac{13}{9}$
$-6\frac{5}{9}$
$\overline{3\frac{8}{9}}$

$9-6=3$, and $\frac{13}{9}-\frac{5}{9}=\frac{8}{9}$.
So, $10\frac{4}{9}-6\frac{5}{9}=\mathbf{3\frac{8}{9}}$.

We check our answer with addition: $\boxed{3\frac{8}{9}}+6\frac{5}{9}=10\frac{4}{9}$. ✓

138. We cannot take away 2 thirds from 1 third. So, we regroup $4\frac{1}{3}$. We take $1=\frac{3}{3}$ from the 4 and add it to $\frac{1}{3}$:
$4\frac{1}{3}=3+\frac{3}{3}+\frac{1}{3}=3+\frac{4}{3}$.
So, $4\frac{1}{3}=3\frac{4}{3}$. Now, we subtract.

$3\frac{4}{3}$
$-2\frac{2}{3}$
$\overline{1\frac{2}{3}}$

$3-2=1$, and $\frac{4}{3}-\frac{2}{3}=\frac{2}{3}$.
So, $4\frac{1}{3}-2\frac{2}{3}=\mathbf{1\frac{2}{3}}$.

139. We cannot take away 9 elevenths from 5 elevenths. So, we regroup $7\frac{5}{11}$. We take $1=\frac{11}{11}$ from the 7 and add it to $\frac{5}{11}$ to get $7\frac{5}{11}=6\frac{16}{11}$. Now, we subtract.

$7\frac{5}{11}$ → $6\frac{16}{11}$
$-3\frac{9}{11}$ $-3\frac{9}{11}$
                  $\overline{3\frac{7}{11}}$

So, $7\frac{5}{11}-3\frac{9}{11}=\mathbf{3\frac{7}{11}}$.

140. We cannot take away 3 fourths from 1 fourth. So, we regroup $8\frac{1}{4}$. We take $1=\frac{4}{4}$ from the 8 and add it to $\frac{1}{4}$ to get $8\frac{1}{4}=7\frac{5}{4}$. Now, we subtract.

$8\frac{1}{4}$ → $7\frac{5}{4}$
$-3\frac{3}{4}$ $-3\frac{3}{4}$
                 $\overline{4\frac{2}{4}}$

So, $8\frac{1}{4}-3\frac{3}{4}=4\frac{2}{4}=\mathbf{4\frac{1}{2}}$.

141. We cannot take away 11 fourteenths from 5 fourteenths. So, we regroup $8\frac{5}{14}$. We take $1=\frac{14}{14}$ from the 8 and add it to $\frac{5}{14}$ to get $8\frac{5}{14}=7\frac{19}{14}$. Now, we can subtract.

$8\frac{5}{14}$ → $7\frac{19}{14}$
$-6\frac{11}{14}$ $-6\frac{11}{14}$
                    $\overline{1\frac{8}{14}}$

So, $8\frac{5}{14}-6\frac{11}{14}=1\frac{8}{14}=\mathbf{1\frac{4}{7}}$.

142. We cannot take away 7 ninths from 1 ninth. So, we regroup $12\frac{1}{9}$. We take $1=\frac{9}{9}$ from the 12 and add it to $\frac{1}{9}$ to get $12\frac{1}{9}=11\frac{10}{9}$. Now, we subtract.

$12\frac{1}{9}$ → $11\frac{10}{9}$
$-5\frac{7}{9}$ $-5\frac{7}{9}$
                    $\overline{6\frac{3}{9}}$

So, $12\frac{1}{9}-5\frac{7}{9}=6\frac{3}{9}=\mathbf{6\frac{1}{3}}$.

FRACTIONS
Mixed Number Arithmetic   62-63

143. We add the integer and fractional parts separately:
$2\frac{16}{17}+\frac{9}{17}=2+\left(\frac{16}{17}+\frac{9}{17}\right)$
$=2+\frac{25}{17}$
$=2+\frac{17}{17}+\frac{8}{17}$
$=2+1+\frac{8}{17}$
$=3+\frac{8}{17}$.

So, $2\frac{16}{17}+\frac{9}{17}=\mathbf{3\frac{8}{17}}$.

— or —

To add nine seventeenths to $2\frac{16}{17}$, we start by adding one seventeenth to $2\frac{16}{17}$ to get a whole number. This gives us $2\frac{16}{17}+\frac{1}{17}=3$. Then, we add the remaining eight seventeenths: $3+\frac{8}{17}=3\frac{8}{17}$.
We write $2\frac{16}{17}+\frac{9}{17}=2\frac{16}{17}+\left(\frac{1}{17}+\frac{8}{17}\right)$
$=\left(2\frac{16}{17}+\frac{1}{17}\right)+\frac{8}{17}$
$=3+\frac{8}{17}$
$=\mathbf{3\frac{8}{17}}$.

144. To add six thirteenths to $4\frac{12}{13}$, we start by adding one thirteenth to $4\frac{12}{13}$ to get a whole number. This gives us $4\frac{12}{13}+\frac{1}{13}=5$. Then, we add the remaining five thirteenths: $5+\frac{5}{13}=5\frac{5}{13}$.
We write $4\frac{12}{13}+\frac{6}{13}=4\frac{12}{13}+\left(\frac{1}{13}+\frac{5}{13}\right)$
$=\left(4\frac{12}{13}+\frac{1}{13}\right)+\frac{5}{13}$
$=5+\frac{5}{13}$
$=\mathbf{5\frac{5}{13}}$.

145. To add $8\frac{7}{23}$ to $9\frac{22}{23}$, we start by adding one twenty-third to $9\frac{22}{23}$ to get a whole number. This gives us $9\frac{22}{23}+\frac{1}{23}=10$. Then, we add the remaining $8\frac{6}{23}$ to get $10+8\frac{6}{23}=18\frac{6}{23}$.

We write $9\frac{22}{23}+8\frac{7}{23} = 9\frac{22}{23}+\left(\frac{1}{23}+8\frac{6}{23}\right)$
$= \left(9\frac{22}{23}+\frac{1}{23}\right)+8\frac{6}{23}$
$= 10+8\frac{6}{23}$
$= \mathbf{18\frac{6}{23}}.$

146. To add $3\frac{12}{19}$ to $7\frac{17}{19}$, we start by adding *two* nineteenths to $7\frac{17}{19}$ to get a whole number. This gives us $7\frac{17}{19}+\frac{2}{19} = 8$.
Then, we add the remaining $3\frac{10}{19}$ to get $8+3\frac{10}{19} = 11\frac{10}{19}$.

We write $7\frac{17}{19}+3\frac{12}{19} = 7\frac{17}{19}+\left(\frac{2}{19}+3\frac{10}{19}\right)$
$= \left(7\frac{17}{19}+\frac{2}{19}\right)+3\frac{10}{19}$
$= 8+3\frac{10}{19}$
$= \mathbf{11\frac{10}{19}}.$

147. To add $5\frac{7}{12}$ to $2\frac{11}{12}$, we start by adding one twelfth to $2\frac{11}{12}$ to get a whole number. This gives us $2\frac{11}{12}+\frac{1}{12} = 3$.
Then, we add the remaining $5\frac{6}{12}$ to get $5\frac{6}{12}+3 = 8\frac{6}{12}$.

We write $5\frac{7}{12}+2\frac{11}{12} = \left(5\frac{6}{12}+\frac{1}{12}\right)+2\frac{11}{12}$
$= 5\frac{6}{12}+\left(\frac{1}{12}+2\frac{11}{12}\right)$
$= 5\frac{6}{12}+3$
$= 8\frac{6}{12}$
$= \mathbf{8\frac{1}{2}}.$

148. To add $9\frac{19}{21}$ to $3\frac{20}{21}$, we start by adding one twenty-first to $3\frac{20}{21}$ to get a whole number. This gives us $3\frac{20}{21}+\frac{1}{21} = 4$.
Then, we add the remaining $9\frac{18}{21}$ to get $9\frac{18}{21}+4 = 13\frac{18}{21}$.

We write $9\frac{19}{21}+3\frac{20}{21} = \left(9\frac{18}{21}+\frac{1}{21}\right)+3\frac{20}{21}$
$= 9\frac{18}{21}+\left(\frac{1}{21}+3\frac{20}{21}\right)$
$= 9\frac{18}{21}+4$
$= 13\frac{18}{21}$
$= \mathbf{13\frac{6}{7}}.$

For the following solutions, we show a couple of different strategies. You may have used other strategies to arrive at the same final answers.

149. We count up from $2\frac{14}{15}$ to $4\frac{7}{15}$. From $2\frac{14}{15}$ to 3 is $\frac{1}{15}$. Then, from 3 to $4\frac{7}{15}$ is $1\frac{7}{15}$ more.

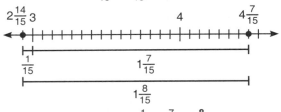

So, the difference equals $\frac{1}{15}+1\frac{7}{15} = \mathbf{1\frac{8}{15}}.$
— *or* —

$2\frac{14}{15}$ is $\frac{1}{15}$ less than 3. If we add $\frac{1}{15}$ to both numbers in a subtraction problem, the difference between the new numbers is the same as the difference between the old numbers. So, we add $\frac{1}{15}$ to both $4\frac{7}{15}$ and $2\frac{14}{15}$ to make the subtraction easier.

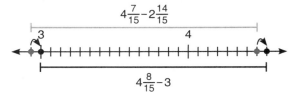

So, $4\frac{7}{15}-2\frac{14}{15}$ is equal to
$\left(4\frac{7}{15}+\frac{1}{15}\right)-\left(2\frac{14}{15}+\frac{1}{15}\right) = 4\frac{8}{15}-3 = \mathbf{1\frac{8}{15}}.$
— *or* —

To subtract $2\frac{14}{15}$, we can take away 3 and give back $\frac{1}{15}$.

$4\frac{7}{15}-2\frac{14}{15} = 4\frac{7}{15}-3+\frac{1}{15}$
$= 1\frac{7}{15}+\frac{1}{15}$
$= \mathbf{1\frac{8}{15}}.$

We check our answer with addition:
$\boxed{1\frac{8}{15}}+2\frac{14}{15} = 4\frac{7}{15}.$ ✓

150. We count up. From $5\frac{10}{11}$ to 6 is $\frac{1}{11}$. Then, from 6 to $9\frac{2}{11}$ is $3\frac{2}{11}$ more.

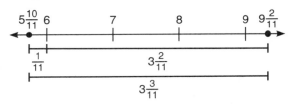

So, the difference equals $\frac{1}{11}+3\frac{2}{11} = \mathbf{3\frac{3}{11}}.$
— *or* —

$5\frac{10}{11}$ is $\frac{1}{11}$ less than 6. We add $\frac{1}{11}$ to both $9\frac{2}{11}$ and $5\frac{10}{11}$ to make the subtraction easier.

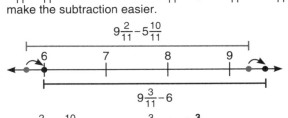

So, $9\frac{2}{11}-5\frac{10}{11}$ is equal to $9\frac{3}{11}-6 = \mathbf{3\frac{3}{11}}.$
— *or* —

To subtract $5\frac{10}{11}$, we can take away 6 and give back $\frac{1}{11}$.

$9\frac{2}{11}-5\frac{10}{11} = 9\frac{2}{11}-6+\frac{1}{11}$
$= 3\frac{2}{11}+\frac{1}{11}$
$= \mathbf{3\frac{3}{11}}.$

151. We count up. From $3\frac{6}{7}$ to 4 is $\frac{1}{7}$. Then, from 4 to $6\frac{4}{7}$ is $2\frac{4}{7}$ more.

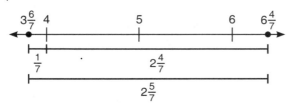

So, the difference equals $\frac{1}{7}+2\frac{4}{7} = \mathbf{2\frac{5}{7}}$.

— or —

$3\frac{6}{7}$ is $\frac{1}{7}$ less than 4. We add $\frac{1}{7}$ to both $6\frac{4}{7}$ and $3\frac{6}{7}$ to make the subtraction easier.

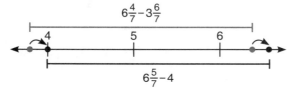

So, $6\frac{4}{7}-3\frac{6}{7}$ is equal to $6\frac{5}{7}-4 = \mathbf{2\frac{5}{7}}$.

— or —

To subtract $3\frac{6}{7}$, we can take away 4 and give back $\frac{1}{7}$.

$6\frac{4}{7}-3\frac{6}{7} = 6\frac{4}{7}-4+\frac{1}{7}$
$= 2\frac{4}{7}+\frac{1}{7}$
$= \mathbf{2\frac{5}{7}}$.

152. We count up. From $3\frac{15}{17}$ to 4 is $\frac{2}{17}$. Then, from 4 to $4\frac{12}{17}$ is $\frac{12}{17}$ more.

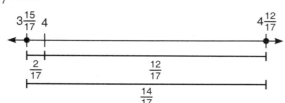

So, the difference equals $\frac{2}{17}+\frac{12}{17} = \mathbf{\frac{14}{17}}$.

— or —

$3\frac{15}{17}$ is $\frac{2}{17}$ less than 4. We add $\frac{2}{17}$ to both $4\frac{12}{17}$ and $3\frac{15}{17}$ to make the subtraction easier.

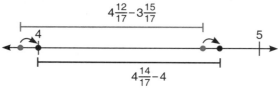

So, $4\frac{12}{17}-3\frac{15}{17}$ is equal to $4\frac{14}{17}-4 = \mathbf{\frac{14}{17}}$.

— or —

To subtract $3\frac{15}{17}$, we can take away 4 and give back $\frac{2}{17}$.

$4\frac{12}{17}-3\frac{15}{17} = 4\frac{12}{17}-4+\frac{2}{17}$
$= \frac{12}{17}+\frac{2}{17}$
$= \mathbf{\frac{14}{17}}$.

153. We count up. From $1\frac{7}{8}$ to 2 is $\frac{1}{8}$. Then, from 2 to $5\frac{3}{8}$ is $3\frac{3}{8}$ more.

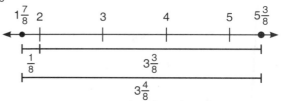

So, the difference equals $\frac{1}{8}+3\frac{3}{8} = 3\frac{4}{8} = \mathbf{3\frac{1}{2}}$.

— or —

$1\frac{7}{8}$ is $\frac{1}{8}$ less than 2. We add $\frac{1}{8}$ to both $5\frac{3}{8}$ and $1\frac{7}{8}$ to make the subtraction easier.

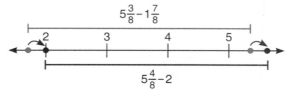

So, $5\frac{3}{8}-1\frac{7}{8}$ is equal to $5\frac{4}{8}-2 = 3\frac{4}{8} = \mathbf{3\frac{1}{2}}$.

— or —

To subtract $1\frac{7}{8}$, we can take away 2 and give back $\frac{1}{8}$.

$5\frac{3}{8}-1\frac{7}{8} = 5\frac{3}{8}-2+\frac{1}{8}$
$= 3\frac{3}{8}+\frac{1}{8}$
$= 3\frac{4}{8}$
$= \mathbf{3\frac{1}{2}}$.

154. We count up. From $2\frac{10}{13}$ to 3 is $\frac{3}{13}$. Then, from 3 to $12\frac{8}{13}$ is $9\frac{8}{13}$ more.

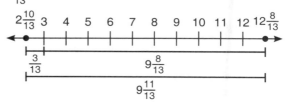

So, the difference equals $\frac{3}{13}+9\frac{8}{13} = \mathbf{9\frac{11}{13}}$.

— or —

$2\frac{10}{13}$ is $\frac{3}{13}$ less than 3. We add $\frac{3}{13}$ to both $12\frac{8}{13}$ and $2\frac{10}{13}$ to make the subtraction easier.

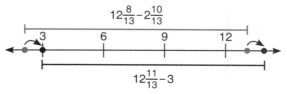

So, $12\frac{8}{13}-2\frac{10}{13}$ is equal to $12\frac{11}{13}-3 = \mathbf{9\frac{11}{13}}$.

— or —

To subtract $2\frac{10}{13}$, we can take away 3 and give back $\frac{3}{13}$.

$$12\frac{8}{13} - 2\frac{10}{13} = 12\frac{8}{13} - 3 + \frac{3}{13}$$
$$= 9\frac{8}{13} + \frac{3}{13}$$
$$= \mathbf{9\frac{11}{13}}.$$

FRACTIONS

Practice 64-65

155. We add the whole numbers and fractions separately.

$$5\frac{1}{3} + 16\frac{1}{3} = (5+16) + \left(\frac{1}{3} + \frac{1}{3}\right)$$
$$= 21 + \frac{2}{3}$$
$$= \mathbf{21\frac{2}{3}}.$$

156. We cannot take away $\frac{7}{10}$ from $\frac{3}{10}$. So, we regroup $7\frac{3}{10}$.

$$\begin{array}{c} 7\frac{3}{10} \\ -2\frac{7}{10} \\ \hline \end{array} \longrightarrow \begin{array}{c} 6\frac{13}{10} \\ -2\frac{7}{10} \\ \hline 4\frac{6}{10} \end{array}$$

So, $7\frac{3}{10} - 2\frac{7}{10} = 4\frac{6}{10} = \mathbf{4\frac{3}{5}}.$

— or —

We add $\frac{3}{10}$ to $7\frac{3}{10}$ and to $2\frac{7}{10}$ to make the subtraction easier: $7\frac{3}{10} - 2\frac{7}{10} = 7\frac{6}{10} - 3 = 4\frac{6}{10} = \mathbf{4\frac{3}{5}}.$

157. We add the whole numbers and fractions separately.

$$67\frac{3}{8} + 14\frac{7}{8} = (67+14) + \left(\frac{3}{8} + \frac{7}{8}\right)$$
$$= 81 + \frac{10}{8}$$
$$= 81 + \frac{8}{8} + \frac{2}{8}$$
$$= 82\frac{2}{8}$$
$$= \mathbf{82\frac{1}{4}}.$$

— or —

To add $67\frac{3}{8}$ to $14\frac{7}{8}$, we begin by adding $\frac{1}{8}$ to $14\frac{7}{8}$ to get a whole number: $14\frac{7}{8} + \frac{1}{8} = 15$.
Then we add the remaining $67\frac{2}{8}$ to get $67\frac{2}{8} + 15 = 82\frac{2}{8}.$

$$67\frac{3}{8} + 14\frac{7}{8} = \left(67\frac{2}{8} + \frac{1}{8}\right) + 14\frac{7}{8}$$
$$= 67\frac{2}{8} + \left(\frac{1}{8} + 14\frac{7}{8}\right)$$
$$= 67\frac{2}{8} + 15$$
$$= 82 + \frac{2}{8}$$
$$= \mathbf{82\frac{1}{4}}.$$

158. $\frac{271}{300} - \frac{201}{300} = \frac{70}{300}$. We simplify to get $\frac{70}{300} = \mathbf{\frac{7}{30}}.$

159. $\frac{27}{50} + \frac{37}{50} = \frac{64}{50}$. We convert to a mixed number in simplest form and get $\frac{64}{50} = \frac{50}{50} + \frac{14}{50} = 1\frac{14}{50} = \mathbf{1\frac{7}{25}}.$

160. We add the whole numbers and fractions separately.
We write $32\frac{7}{11} + 84\frac{9}{11} = (32+84) + \left(\frac{7}{11} + \frac{9}{11}\right)$
$$= 116 + \frac{16}{11}$$
$$= 116 + \frac{11}{11} + \frac{5}{11}$$
$$= \mathbf{117\frac{5}{11}}.$$

— or —

To add $32\frac{7}{11}$ to $84\frac{9}{11}$, we begin by adding $\frac{2}{11}$ to $84\frac{9}{11}$ to get a whole number: $84\frac{9}{11} + \frac{2}{11} = 85$.
Then, we add the remaining $32\frac{5}{11}$ to get $32\frac{5}{11} + 85 = 117\frac{5}{11}.$

We write $32\frac{7}{11} + 84\frac{9}{11} = \left(32\frac{5}{11} + \frac{2}{11}\right) + 84\frac{9}{11}$
$$= 32\frac{5}{11} + \left(\frac{2}{11} + 84\frac{9}{11}\right)$$
$$= 32\frac{5}{11} + 85$$
$$= 117 + \frac{5}{11}$$
$$= \mathbf{117\frac{5}{11}}.$$

161. We cannot take away $\frac{13}{18}$ from $\frac{1}{18}$. So, we regroup $17\frac{1}{18}$.

$$\begin{array}{c} 17\frac{1}{18} \\ -5\frac{13}{18} \\ \hline \end{array} \longrightarrow \begin{array}{c} 16\frac{19}{18} \\ -5\frac{13}{18} \\ \hline 11\frac{6}{18} \end{array}$$

So, $17\frac{1}{18} - 5\frac{13}{18} = 11\frac{6}{18} = \mathbf{11\frac{1}{3}}.$

162. We count up. From $33\frac{18}{19}$ to 34 is $\frac{1}{19}$. Then, from 34 to 74 is 40 more.

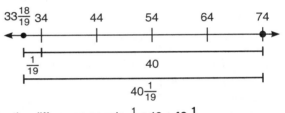

So, the difference equals $\frac{1}{19} + 40 = \mathbf{40\frac{1}{19}}.$

— or —

74 is a whole number, so $74 = 74\frac{0}{19}$. We cannot take away $\frac{18}{19}$ from $\frac{0}{19}$, so we regroup 74.

$$\begin{array}{c} 74 \\ -33\frac{18}{19} \\ \hline \end{array} \longrightarrow \begin{array}{c} 73\frac{19}{19} \\ -33\frac{18}{19} \\ \hline 40\frac{1}{19} \end{array}$$

Therefore, $74 - 33\frac{18}{19} = \mathbf{40\frac{1}{19}}.$

163. We subtract the whole numbers and fractions separately.

$$\begin{array}{r} 526\frac{7}{8} \\ -213\frac{3}{8} \\ \hline 313\frac{4}{8} \end{array}$$

So, $526\frac{7}{8} - 213\frac{3}{8} = 313\frac{4}{8} = \mathbf{313\frac{1}{2}}.$

164. We add the whole numbers and fractions separately.

$$421\frac{4}{9} + 82\frac{1}{9} = (421+82) + \left(\frac{4}{9} + \frac{1}{9}\right)$$
$$= 503 + \frac{5}{9}$$
$$= \mathbf{503\frac{5}{9}}.$$

165. We add and subtract from left to right.
$$12\tfrac{2}{9}+12\tfrac{4}{9}-5\tfrac{2}{9}=\left(12\tfrac{2}{9}+12\tfrac{4}{9}\right)-5\tfrac{2}{9}$$
$$=24\tfrac{6}{9}-5\tfrac{2}{9}$$
$$=\mathbf{19\tfrac{4}{9}}.$$

166. We add and subtract from left to right.
$$20\tfrac{1}{3}+10\tfrac{1}{3}-10\tfrac{2}{3}-5\tfrac{2}{3}=30\tfrac{2}{3}-10\tfrac{2}{3}-5\tfrac{2}{3}$$
$$=20-5\tfrac{2}{3}$$
$$=19\tfrac{3}{3}-5\tfrac{2}{3}$$
$$=\mathbf{14\tfrac{1}{3}}.$$

167. We subtract $10\tfrac{1}{4}-3\tfrac{3}{4}$ to determine the number of ounces of juice Anna originally had in her mug.

$$\begin{array}{r}10\tfrac{1}{4}\\-3\tfrac{3}{4}\\\hline\end{array}\rightarrow\begin{array}{r}9\tfrac{5}{4}\\-3\tfrac{3}{4}\\\hline 6\tfrac{2}{4}\end{array}$$

So, Anna had $6\tfrac{2}{4}=\mathbf{6\tfrac{1}{2}}$ **ounces** of juice in her mug.

— or —

$3\tfrac{3}{4}$ is $\tfrac{1}{4}$ less than 4. We add $\tfrac{1}{4}$ to $10\tfrac{1}{4}$ and to $3\tfrac{3}{4}$ to make the subtraction easier.

So, $10\tfrac{1}{4}-3\tfrac{3}{4}=10\tfrac{2}{4}-4=6\tfrac{2}{4}=6\tfrac{1}{2}$.

Anna had **$6\tfrac{1}{2}$ ounces** of juice in her mug.

168. We begin by finding the length of Grogg's hair. Nellie's hair is 23 inches long, which is $18\tfrac{1}{3}$ inches longer than Grogg's hair. So, Grogg's hair is $23-18\tfrac{1}{3}=22\tfrac{3}{3}-18\tfrac{1}{3}=4\tfrac{2}{3}$ inches long.

Fiona's hair is $31\tfrac{2}{3}$ inches longer than Grogg's hair. So, Fiona's hair is $4\tfrac{2}{3}+31\tfrac{2}{3}=35+\tfrac{4}{3}=\mathbf{36\tfrac{1}{3}}$ **inches** long.

— or —

Fiona's hair is $31\tfrac{2}{3}$ inches longer than Grogg's hair, and Nellie's hair is $18\tfrac{1}{3}$ inches longer than Grogg's hair.

So, Fiona's hair is $31\tfrac{2}{3}-18\tfrac{1}{3}=13\tfrac{1}{3}$ inches longer than Nellie's hair.

Since Nellie's hair is 23 inches long, Fiona's hair is $23+13\tfrac{1}{3}=\mathbf{36\tfrac{1}{3}}$ **inches** long.

169. We draw a diagram. The sum of both heights is $82\tfrac{1}{4}$ in.

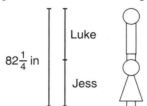

Luke is $5\tfrac{1}{4}$ inches taller than Jess. So, in our diagram, we can replace Luke with Jess plus $5\tfrac{1}{4}$ inches.

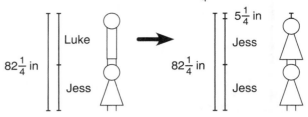

So, twice Jess's height is $82\tfrac{1}{4}-5\tfrac{1}{4}=77$ inches.

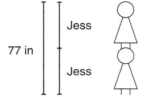

Therefore, Jess is $77\div 2=\tfrac{77}{2}=\tfrac{76}{2}+\tfrac{1}{2}=\mathbf{38\tfrac{1}{2}}$ **inches** tall.

— or —

We use j to represent Jess's height.

Luke is $5\tfrac{1}{4}$ inches taller than Jess, so we write an expression to represent Luke's height: $j+5\tfrac{1}{4}$.

The sum of their heights is $j+\left(j+5\tfrac{1}{4}\right)=82\tfrac{1}{4}$.

We subtract $5\tfrac{1}{4}$ from both sides of this equation to get
$$j+\left(j+5\tfrac{1}{4}\right)=82\tfrac{1}{4}$$
$$j+j+5\tfrac{1}{4}-5\tfrac{1}{4}=82\tfrac{1}{4}-5\tfrac{1}{4}$$
$$j+j=77$$
$$2\times j=77.$$

Then, to find the number that we multiply by 2 to get 77, we divide 77 by 2:
$$j=77\div 2=\tfrac{77}{2}=\tfrac{76}{2}+\tfrac{1}{2}=38\tfrac{1}{2}.$$

So, Jess is **$38\tfrac{1}{2}$ inches** tall.

170. Patrice made $19\tfrac{3}{5}$ ounces of fudge, so Michael made $19\tfrac{3}{5}+9\tfrac{4}{5}=28+\tfrac{7}{5}=29\tfrac{2}{5}$ ounces of fudge.

Together, Patrice and Michael made $19\tfrac{3}{5}+29\tfrac{2}{5}=48+\tfrac{5}{5}=49$ ounces of fudge.

When they divide this fudge equally into 35 bags, each bag has $49\div 35=\tfrac{49}{35}=\tfrac{35}{35}+\tfrac{14}{35}=1+\tfrac{14}{35}=\mathbf{1\tfrac{2}{5}}$ **ounces**.

171. $1\tfrac{1}{5}+1\tfrac{4}{5}=3.$
$\tfrac{3}{5}+2\tfrac{2}{5}=3.$
$\tfrac{1}{5}+2\tfrac{4}{5}=3.$
$2\tfrac{1}{5}+\tfrac{4}{5}=3.$
$1\tfrac{2}{5}+1\tfrac{3}{5}=3.$

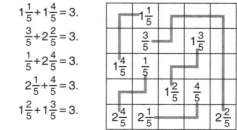

172. $1\frac{3}{7}+2\frac{4}{7}=4$.
$3\frac{3}{7}+\frac{4}{7}=4$.
$2\frac{6}{7}+1\frac{1}{7}=4$.
$3\frac{5}{7}+\frac{2}{7}=4$.

173. $2\frac{2}{5}+2\frac{3}{5}=5$.
$1\frac{4}{5}+3\frac{1}{5}=5$.
$2\frac{4}{5}+2\frac{1}{5}=5$.
$\frac{2}{5}+4\frac{3}{5}=5$.

174. $1\frac{4}{9}+4\frac{5}{9}=6$.
$4\frac{8}{9}+1\frac{1}{9}=6$.
$2\frac{4}{9}+3\frac{5}{9}=6$.
$5\frac{7}{9}+\frac{2}{9}=6$.

175. $1\frac{4}{9}+3\frac{5}{9}=5$.
$1\frac{1}{9}+3\frac{8}{9}=5$.
$2\frac{4}{9}+2\frac{5}{9}=5$.
$\frac{2}{9}+4\frac{7}{9}=5$.
$1\frac{8}{9}+3\frac{1}{9}=5$.
$2\frac{7}{9}+2\frac{2}{9}=5$.
$\frac{5}{9}+4\frac{4}{9}=5$.

176. $3\frac{2}{3}+3\frac{1}{3}=7$.
$2\frac{2}{3}+4\frac{1}{3}=7$.
$6\frac{1}{3}+\frac{2}{3}=7$.
$2\frac{1}{3}+4\frac{2}{3}=7$.
$5\frac{1}{3}+1\frac{2}{3}=7$.

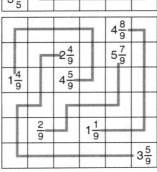
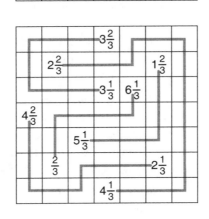

FRACTIONS
Longer Sums 68–69

177. We add the numerators from left to right, put the result over 23, and then write the sum as a mixed number:

$\frac{11}{23}+\frac{16}{23}+\frac{12}{23}=\frac{39}{23}=\frac{23}{23}+\frac{16}{23}=\mathbf{1\frac{16}{23}}$.

— or —

We notice that there is one pair of fractions whose sum is 1.

$\frac{11}{23}+\frac{16}{23}+\frac{12}{23}$

So, we reorder and regroup the addition to make this sum easier to compute.

$\frac{11}{23}+\frac{16}{23}+\frac{12}{23}=\left(\frac{11}{23}+\frac{12}{23}\right)+\frac{16}{23}$
$=1+\frac{16}{23}$
$=\mathbf{1\frac{16}{23}}$.

178. We notice that there is one pair of fractions whose sum is 1.

$\frac{9}{11}+\frac{1}{11}+\frac{10}{11}$

So, we regroup the addition to make this sum easier to compute.

$\frac{9}{11}+\frac{1}{11}+\frac{10}{11}=\frac{9}{11}+\left(\frac{1}{11}+\frac{10}{11}\right)$
$=\frac{9}{11}+1$
$=\mathbf{1\frac{9}{11}}$.

179. We notice that there is one pair of mixed numbers whose sum is a whole number.

$2\frac{2}{15}+\frac{8}{15}+1\frac{13}{15}$

So, we reorder and regroup the addition to make this sum easier to compute.

$2\frac{2}{15}+\frac{8}{15}+1\frac{13}{15}=\left(2\frac{2}{15}+1\frac{13}{15}\right)+\frac{8}{15}$
$=4+\frac{8}{15}$
$=\mathbf{4\frac{8}{15}}$.

180. We notice that there are three pairs of fractions whose sum is 1.

$\frac{1}{7}+\frac{2}{7}+\frac{3}{7}+\frac{4}{7}+\frac{5}{7}+\frac{6}{7}$

So, we reorder and regroup the addition to make this sum easier to compute.

$\frac{1}{7}+\frac{2}{7}+\frac{3}{7}+\frac{4}{7}+\frac{5}{7}+\frac{6}{7}=\left(\frac{1}{7}+\frac{6}{7}\right)+\left(\frac{2}{7}+\frac{5}{7}\right)+\left(\frac{3}{7}+\frac{4}{7}\right)$
$=1+1+1$
$=\mathbf{3}$.

181. The perimeter of the triangle is $5\frac{5}{8}+7\frac{7}{8}+3\frac{3}{8}$ inches.
We notice that there is one pair of mixed numbers whose sum is a whole number, so we reorder and group this pair to make this sum easier to compute.

$$5\frac{5}{8}+7\frac{7}{8}+3\frac{3}{8} = \left(5\frac{5}{8}+3\frac{3}{8}\right)+7\frac{7}{8}$$
$$= 9+7\frac{7}{8}$$
$$= 16\frac{7}{8}.$$

So, the perimeter of the triangle is $16\frac{7}{8}$ **inches**.

182. The perimeter of the rectangle is $5\frac{7}{10}+2\frac{3}{10}+5\frac{7}{10}+2\frac{3}{10}$ meters. We notice that there are two pairs of mixed numbers whose sum is 8, so we group these pairs to make this sum easier to compute.

$$5\frac{7}{10}+2\frac{3}{10}+5\frac{7}{10}+2\frac{3}{10} = \left(5\frac{7}{10}+2\frac{3}{10}\right)+\left(5\frac{7}{10}+2\frac{3}{10}\right)$$
$$= 8+8$$
$$= 16.$$

So, the perimeter of the rectangle is **16 meters**.

183. The perimeter of the pentagon is $5\frac{1}{5}+5\frac{3}{5}+5\frac{2}{5}+5\frac{4}{5}+5\frac{2}{5}$ inches. We notice that there are two pairs of mixed numbers whose sum is a whole number, so we reorder and regroup to make this sum easier to compute.

$$5\frac{1}{5}+5\frac{3}{5}+5\frac{2}{5}+5\frac{4}{5}+5\frac{2}{5} = \left(5\frac{1}{5}+5\frac{4}{5}\right)+\left(5\frac{2}{5}+5\frac{3}{5}\right)+5\frac{2}{5}$$
$$= 11+11+5\frac{2}{5}$$
$$= 22+5\frac{2}{5}$$
$$= 27\frac{2}{5}.$$

So, the perimeter of the pentagon is $27\frac{2}{5}$ **inches**.

184. Since this shape is rectilinear, the length of its left side is given by the sum of the two side lengths on the right: $3\frac{2}{3}+4\frac{2}{3}=7\frac{4}{3}=8\frac{1}{3}$.

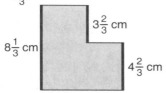

Similarly, the length of the bottom side is given by the sum of the two side lengths along the top: $4\frac{1}{3}+3\frac{1}{3}=7\frac{2}{3}$.

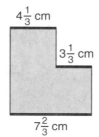

Now we know the lengths of each of the 6 sides, and we add. The perimeter of the shape is

$$3\frac{2}{3}+4\frac{2}{3}+8\frac{1}{3}+4\frac{1}{3}+3\frac{1}{3}+7\frac{2}{3}$$
$$= \left(3\frac{2}{3}+4\frac{1}{3}\right)+\left(4\frac{2}{3}+3\frac{1}{3}\right)+\left(8\frac{1}{3}+7\frac{2}{3}\right)$$
$$= 8+8+16$$
$$= \textbf{32 cm}.$$

— or —

The perimeter of this shape is the same as the perimeter of a $8\frac{1}{3}$ cm by $7\frac{2}{3}$ cm rectangle:

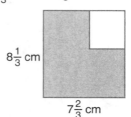

So, the perimeter is $8\frac{1}{3}+7\frac{2}{3}+8\frac{1}{3}+7\frac{2}{3}$ centimeters. We notice there are two pairs of mixed numbers whose sum is 16, so we group these pairs to make this sum easier to compute.

$$8\frac{1}{3}+7\frac{2}{3}+8\frac{1}{3}+7\frac{2}{3} = \left(8\frac{1}{3}+7\frac{2}{3}\right)+\left(8\frac{1}{3}+7\frac{2}{3}\right)$$
$$= 16+16$$
$$= 32.$$

So, the perimeter of the rectilinear hexagon is **32 cm**.

Challenge Problems   70-71

185. We must start with one block. We continue to add one block to the stack and record the height until we find a height that is a whole number.

1 block: $2\frac{3}{5}$ inches.

2 blocks: $2\frac{3}{5}+2\frac{3}{5}=4+\frac{6}{5}=5\frac{1}{5}$ inches.

3 blocks: $5\frac{1}{5}+2\frac{3}{5}=7\frac{4}{5}$ inches.

4 blocks: $7\frac{4}{5}+2\frac{3}{5}=9+\frac{7}{5}=10\frac{2}{5}$ inches.

5 blocks: $10\frac{2}{5}+2\frac{3}{5}=12+\frac{5}{5}=13$ inches.

So, we need at least **5** blocks to make a stack whose height is a whole number of inches.

— or —

A fraction is equal to a whole number when the numerator is a multiple of the denominator.

Each block adds $2\frac{3}{5}=\frac{13}{5}$ inches to the height of the stack. So, we skip-count by $\frac{13}{5}$'s until we reach a fraction whose numerator is a multiple of 5.

$$\frac{13}{5}, \frac{26}{5}, \frac{39}{5}, \frac{52}{5}, \frac{65}{5}.$$

65 is the smallest multiple of 5 that appears in the numerator. So, the smallest stack whose height is a whole number of inches is $\frac{65}{5}$ inches tall.

To get this height, we must add four blocks on top of the first. Therefore, we need at least **5** blocks.

186. We first evaluate all of the grouped expressions. To make our subtraction easier, we notice that $3-\frac{1}{8}$ is one eighth less than 3. Then, each expression is one eighth less than the expression to its left.

$$\left(3-\frac{1}{8}\right)+\left(3-\frac{2}{8}\right)+\left(3-\frac{3}{8}\right)+\left(3-\frac{4}{8}\right)+\left(3-\frac{5}{8}\right)+\left(3-\frac{6}{8}\right)+\left(3-\frac{7}{8}\right)$$
$$= 2\frac{7}{8}+2\frac{6}{8}+2\frac{5}{8}+2\frac{4}{8}+2\frac{3}{8}+2\frac{2}{8}+2\frac{1}{8}.$$

Next, we notice there are three pairs of mixed numbers whose sum is 5:

$$2\tfrac{7}{8}+2\tfrac{6}{8}+2\tfrac{5}{8}+2\tfrac{4}{8}+2\tfrac{3}{8}+2\tfrac{2}{8}+2\tfrac{1}{8}.$$

So, we group these pairs to make the sum easier to compute.

$$\left(3-\tfrac{1}{8}\right)+\left(3-\tfrac{2}{8}\right)+\left(3-\tfrac{3}{8}\right)+\left(3-\tfrac{4}{8}\right)+\left(3-\tfrac{5}{8}\right)+\left(3-\tfrac{6}{8}\right)+\left(3-\tfrac{7}{8}\right)$$
$$=2\tfrac{7}{8}+2\tfrac{6}{8}+2\tfrac{5}{8}+2\tfrac{4}{8}+2\tfrac{3}{8}+2\tfrac{2}{8}+2\tfrac{1}{8}$$
$$=\left(2\tfrac{7}{8}+2\tfrac{1}{8}\right)+\left(2\tfrac{6}{8}+2\tfrac{2}{8}\right)+\left(2\tfrac{5}{8}+2\tfrac{3}{8}\right)+2\tfrac{4}{8}$$
$$=5+5+5+2\tfrac{4}{8}$$
$$=17\tfrac{4}{8}$$
$$=\mathbf{17\tfrac{1}{2}}.$$

187. On the first day, Yuan runs $\tfrac{5}{6}$ miles.

The second day, she runs $\tfrac{1}{6}$ miles farther than she did on the first day: $\tfrac{5}{6}+\tfrac{1}{6}=\tfrac{6}{6}$ miles.

On the third day, she runs $\tfrac{1}{6}$ miles farther than she did the second day: $\tfrac{6}{6}+\tfrac{1}{6}=\tfrac{7}{6}$ miles.

Similarly, she runs $\tfrac{8}{6}$ miles on the fourth day, $\tfrac{9}{6}$ miles on the fifth day, and so on until the tenth day when she runs $\tfrac{14}{6}$ miles. All together, Yuan runs

$$\tfrac{5}{6}+\tfrac{6}{6}+\tfrac{7}{6}+\tfrac{8}{6}+\tfrac{9}{6}+\tfrac{10}{6}+\tfrac{11}{6}+\tfrac{12}{6}+\tfrac{13}{6}+\tfrac{14}{6}\text{ miles}.$$

We notice that there are four pairs of fractions whose sum is $\tfrac{18}{6}=3$.

$$\tfrac{5}{6}+\tfrac{6}{6}+\tfrac{7}{6}+\tfrac{8}{6}+\tfrac{9}{6}+\tfrac{10}{6}+\tfrac{11}{6}+\tfrac{12}{6}+\tfrac{13}{6}+\tfrac{14}{6}$$

So we group these pairs to make the sum easier to compute.

$$\tfrac{5}{6}+\tfrac{6}{6}+\tfrac{7}{6}+\tfrac{8}{6}+\tfrac{9}{6}+\tfrac{10}{6}+\tfrac{11}{6}+\tfrac{12}{6}+\tfrac{13}{6}+\tfrac{14}{6}$$
$$=\left(\tfrac{5}{6}+\tfrac{13}{6}\right)+\left(\tfrac{6}{6}+\tfrac{12}{6}\right)+\left(\tfrac{7}{6}+\tfrac{11}{6}\right)+\left(\tfrac{8}{6}+\tfrac{10}{6}\right)+\tfrac{9}{6}+\tfrac{14}{6}$$
$$=3+3+3+3+\tfrac{23}{6}$$
$$=12+\tfrac{18}{6}+\tfrac{5}{6}$$
$$=12+3+\tfrac{5}{6}$$
$$=15\tfrac{5}{6}.$$

So, during those ten days, Yuan runs $\mathbf{15\tfrac{5}{6}}$ miles.

188. We let h represent the height of the rectangle. Since the width of the rectangle is twice the height, we know the width is $h+h$.

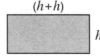

So, the perimeter of the rectangle in centimeters is $(h+h)+h+(h+h)+h=6\times h=21$.

To find the number that we multiply by 6 to get 21, we divide 21 by 6:
$$h=21\div 6=\tfrac{21}{6}=\tfrac{7}{2}=3\tfrac{1}{2}.$$

Therefore, the height of the rectangle is $\mathbf{3\tfrac{1}{2}}$ centimeters.

We check our answer:

If the height of the rectangle is $3\tfrac{1}{2}$ cm, then the width of the rectangle is $3\tfrac{1}{2}+3\tfrac{1}{2}=7$ cm, and the perimeter is
$$\left(3\tfrac{1}{2}+7\right)+\left(3\tfrac{1}{2}+7\right)=10\tfrac{1}{2}+10\tfrac{1}{2}=21\text{ cm. }\checkmark$$

189. The perimeter of the rhombus is made of four equal sides of the triangles. Since the perimeter of each triangle is 8 inches, the length of one side of a triangle is $8\div 3=\tfrac{8}{3}=2\tfrac{2}{3}$ inches.

Therefore, the perimeter of the rhombus is
$$2\tfrac{2}{3}+2\tfrac{2}{3}+2\tfrac{2}{3}+2\tfrac{2}{3}=8+\tfrac{8}{3}$$
$$=8+\tfrac{6}{3}+\tfrac{2}{3}$$
$$=8+2+\tfrac{2}{3}$$
$$=\mathbf{10\tfrac{2}{3}}\text{ inches}.$$

— or —

The perimeter of the rhombus is made of four equal sides of the triangles. Adding three of the triangle side lengths gives the perimeter of a triangle: 8 inches.

We add the fourth equal side length to get the perimeter of the rhombus.

As above, each side has length $8\div 3=\tfrac{8}{3}=2\tfrac{2}{3}$ inches.

Therefore, the perimeter of the rhombus is
$$8+2\tfrac{2}{3}=\mathbf{10\tfrac{2}{3}}\text{ inches}.$$

190. The first cup is $5\tfrac{3}{4}$ inches tall. A 2-cup stack is $8\tfrac{1}{4}$ inches tall, so each extra cup adds $8\tfrac{1}{4}-5\tfrac{3}{4}=2\tfrac{2}{4}$ inches to the height of the stack.

To make a stack of 5 cups, we add 4 cups on top of the first, as shown.

So, the height of the 5-cup stack is

$$5\tfrac{3}{4}+2\tfrac{2}{4}+2\tfrac{2}{4}+2\tfrac{2}{4}+2\tfrac{2}{4}=(5+2+2+2+2)+\tfrac{3}{4}+\left(\tfrac{2}{4}+\tfrac{2}{4}+\tfrac{2}{4}+\tfrac{2}{4}\right)$$
$$=13+\tfrac{3}{4}+\tfrac{8}{4}$$
$$=13+\tfrac{3}{4}+2$$
$$=\mathbf{15\tfrac{3}{4}}\text{ inches}.$$

— or —

Since each extra cup adds $8\tfrac{1}{4}-5\tfrac{3}{4}=2\tfrac{2}{4}$ inches to the height of the stack, we can begin at $5\tfrac{3}{4}$ and skip-count by $2\tfrac{2}{4}$'s.

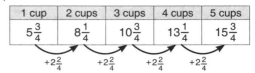

So, the height of the 5-cup stack is $\mathbf{15\tfrac{3}{4}}$ **inches**.

191. A stack of two chairs is $36\frac{7}{8}$ inches tall, and a stack of three chairs is $44\frac{3}{8}$ inches tall. So, placing each additional chair adds $44\frac{3}{8} - 36\frac{7}{8} = 7\frac{4}{8}$ inches to the height of the stack.

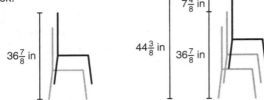

So, the height of one chair is $7\frac{4}{8}$ inches less than the height of the two-chair stack: $36\frac{7}{8} - 7\frac{4}{8} = \mathbf{29\frac{3}{8}}$ **inches**.

We check that $29\frac{3}{8} + 7\frac{4}{8} = 36\frac{7}{8}$ and $29\frac{3}{8} + 7\frac{4}{8} + 7\frac{4}{8} = 44\frac{3}{8}$. ✓

192. The weight of the milk that Joey drank is
$11 - 6\frac{3}{8} = 4\frac{5}{8}$ ounces.

Joey drank half the milk in his full glass, so the weight of the milk in his full glass was
$$4\frac{5}{8} + 4\frac{5}{8} = 8\frac{10}{8} = 9\frac{2}{8} = 9\frac{1}{4} \text{ ounces.}$$

The full glass of milk weighed 11 ounces, so Joey's glass weighs $11 - 9\frac{1}{4} = \mathbf{1\frac{3}{4}}$ **ounces**.

— *or* —

The remaining milk weighs the same amount as the milk that Joey drank: $4\frac{5}{8}$ ounces. If Joey drinks the remaining milk, then the weight of the glass alone is
$6\frac{3}{8} - 4\frac{5}{8} = \mathbf{1\frac{3}{4}}$ **ounces**.

Chapter 9: Solutions

INTEGERS
Negative Numbers — page 73

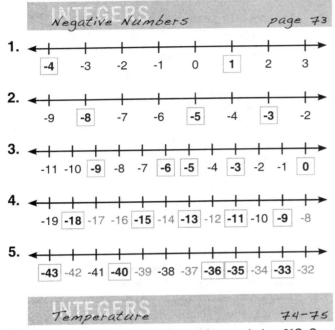

INTEGERS
Temperature — pages 74-75

6. The indicated temperature is two degrees below 0°C. So, the temperature is **-2°C**.

7. The indicated temperature is two degrees below -10°C. So, the temperature is **-12°C**.

8. The indicated temperature is three degrees below -15°C. So, the temperature is **-18°C**.

9. The indicated temperature is three degrees below -30°C. So, the temperature is **-33°C**.

10. The indicated temperature is eight degrees above 0°F. So, the temperature is **8°F**.

11. The indicated temperature is three degrees below 0°F. So, the temperature is **-3°F**.

12. The indicated temperature is three degrees below -20°F. So, the temperature is **-23°F**.

13. The indicated temperature is six degrees below -30°F. So, the temperature is **-36°F**.

14. We locate 10°F on a thermometer and count down 15 degrees.

    The temperature that is 15 degrees colder than 10°F is **-5°F**.

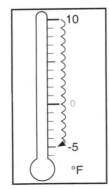

15. We locate -1°C and -13°C on a thermometer. We count 12 degrees from -1°C to -13°C.

    So, the temperature at the park dropped **12 degrees** from afternoon to evening.

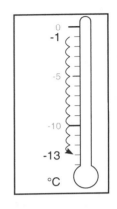

16. We locate these temperatures on a thermometer.

    The coldest temperatures are near the bottom of a thermometer, and the temperatures get warmer as we move up the thermometer.

    So, ordering these temperatures as they appear on the thermometer from bottom to top gives us their order from coldest to warmest: **-29°F, -19°F, -9°F, and 9°F**.

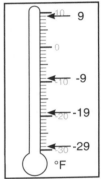

17. We locate -9°C on a thermometer and count up 13 degrees.

    The temperature that is 13 degrees warmer than -9°C is 4°C.

    So, the morning temperature at Beast Academy was **4°C**.

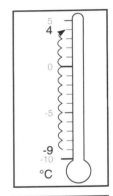

18. The temperature *rose* 22 degrees between midnight and 6 am. So, to find the temperature at midnight, we count *down* 22 degrees from 10°C.

    So, the midnight temperature at Lake Jackalope was **-12°C**.

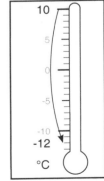

19. We locate -2°C on a thermometer and count down 35 degrees.

    The temperature that is 35 degrees colder than -2°C is -37°C.

    So, the temperature at the top of Mount Everbeast is **-37°C**.

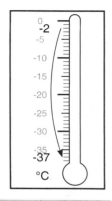

Positive & Negative Integers 76-77

20. The positive integers less than 7 are 1, 2, 3, 4, 5, and 6.

    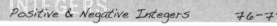

    There are **6** positive integers less than 7.

21. The negative integers greater than -9 are -8, -7, -6, -5, -4, -3, -2, and -1.

    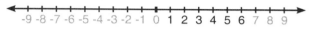

    There are **8** negative integers greater than -9.

22. The integers between -5 and 5 are -4, -3, -2, -1, 0, 1, 2, 3, and 4.

    There are **9** integers less than 5 but greater than -5.

23. There is one integer 3 units to the left of 0 and one integer 3 units to the right of 0.

    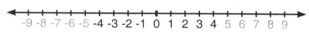

    The two integers that are 3 units from 0 are **-3 and 3**.

24. There is one integer 5 units to the left of -2 and one integer 5 units to the right of -2.

    The two integers that are 5 units from -2 are **-7 and 3**.

25. We first locate -3 and 6 on the number line, then count the distance from one to the other.

    We count that -3 is **9** units from 6.

    — or —

    We know -3 is 3 units to the left of 0, and 6 is 6 units to the right of 0. So, if we start at -3, then we move 3+6 = **9** units right to reach 6.

26. Nonnegative integers are all of the positive integers and zero. So, **0** is the nonnegative integer that is not positive.

27. We look to the number line.

    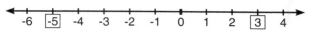

    Since 3 is to the right of -5 on the number line, 3 is greater than -5. Therefore, 3 **>** -5.

    — or —

    We know 3 is greater than 0, while -5 is less than 0. So, 3 is greater than -5. Therefore, 3 **>** -5.

    Any positive number is greater than any negative number.

28. All negative numbers are less than 0. So, -4 **<** 0.

29. We know -5 is 5 units less than 0, while -4 is 4 units less than 0. So, -5 is to the left of -4 on the number line.

    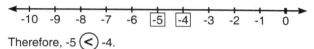

    Therefore, -5 **<** -4.

30. We know -3 is 3 units less than 0, while -5 is 5 units less than 0. So, -3 is to the right of -5 on the number line.

    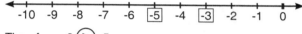

    Therefore, -3 **>** -5.

31. We know -25 is 25 units less than 0, while -18 is 18 units less than 0. So, -25 is to the left of -18 on the number line.

    Therefore, -25 **<** -18.

32. We know -412 is 412 units less than 0, while -1,032 is 1,032 units less than 0. So, -412 is to the right of -1,032 on the number line.

    Therefore, -412 **>** -1,032.

33. We locate these numbers on the number line.

    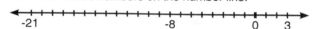

    Ordering these numbers as they appear from left to right gives us their order from least to greatest: **-21 < -8 < 0 < 3**.

34. We locate these numbers on the number line.

    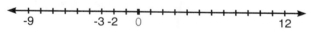

    Ordering these numbers as they appear from left to right gives us their order from least to greatest: **-9 < -3 < -2 < 12**.

35. We locate these numbers on the number line.

    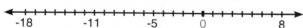

    Ordering these numbers as they appear from left to right gives us their order from least to greatest: **-18 < -11 < -5 < 8**.

36. We locate these numbers on the number line.

    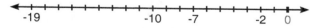

    Ordering these numbers as they appear from left to right gives us their order from least to greatest: **-19 < -10 < -7 < -2**.

INTEGERS
Integer Path Puzzles   78-79

In each of the following puzzles, we begin by circling the smallest number on the grid. We move from hexagon to hexagon, always connecting to the next-smallest integer in the grid. We finish at the largest number on the grid.

37. 38.

39. 40.

41. 42.

43. 44.

45. 46.

INTEGERS
Adding Integers   80-81

47. To add -9+4, we start at -9 and move 4 units to the right.

We arrive at -5. So, -9+4 = **-5**.

— *or* —

We start at -9 and move 4 units to the right. This brings us 4 units closer to zero. We end up 9−4 = 5 units left of zero at **-5**.

48. To add -1+7, we start at -1 and move 7 units to the right.

We arrive at 6. So, -1+7 = **6**.

— *or* —

We start at -1 and move 7 units to the right. We first move 1 unit to the right to arrive at 0, and then we move 7−1 = 6 more units to the right. We end up 6 units to the right of zero at **6**.

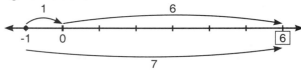

We use the approaches described in the previous solutions to compute the following sums.

49. -7+7 = **0**.   **50.** -3+5 = **2**.
51. -8+13 = **5**.   **52.** -6+2 = **-4**.
53. -8+7 = **-1**.   **54.** -9+1 = **-8**.
55. -1+5 = **4**.   **56.** -8+4 = **-4**.

57. To add 6+(-4), we start at 6 and move 4 units to the left.

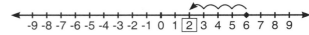

We arrive at 2. So, 6+(-4) = **2**.

— *or* —

We start at 6 and move 4 units to the left. This brings us 4 units closer to zero. We end up 6−4 = 2 units right of zero at **2**.

58. To add -1+(-7), we start at -1 and move 7 units to the left.

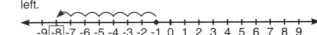

We arrive at -8. So, -1+(-7) = **-8**.

— *or* —

We start at -1 and move 7 units to the left. This brings us 7 units farther from zero. We end up 1+7 = 8 units left of zero at **-8**.

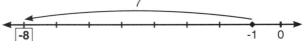

59. To add 2+(-6), we start at 2 and move 6 units to the left.

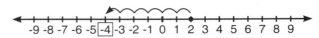

We arrive at -4. So, 2+(-6) = **-4**.

— *or* —

We start at 2 and move 6 units to the left. We first move 2 units to the left to arrive at 0, and then we move 6−2 = 4 more units to the left. We end up 4 units left of zero at **-4**.

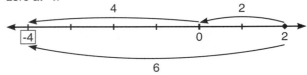

We use the approaches described in the previous solutions to compute the following sums.

60. $-2+(-3) = \mathbf{-5}$.

61. $9+(-3) = \mathbf{6}$.

62. $-3+(-1) = \mathbf{-4}$.

63. To add $8+(-11)+2$, we start at 8, move 11 units to the left, and then move 2 units to the right.

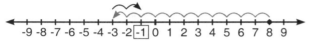

We arrive at -1. So, $8+(-11)+2 = \mathbf{-1}$.

— *or* —

Since addition is commutative and associative, we can add these numbers in any order!

So, $8+(-11)+2 = 8+2+(-11) = 10+(-11) = -1$.

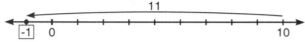

Therefore, $8+(-11)+2 = \mathbf{-1}$.

64. To add $-6+3+(-4)$, we start at -6, move 3 units to the right, and then move 4 units to the left.

We arrive at -7. So, $-6+3+(-4) = \mathbf{-7}$.

— *or* —

$-6+3+(-4) = -6+(-4)+3 = -10+3 = \mathbf{-7}$.

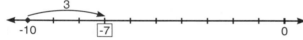

65. To add $5+(-4)+(-1)$, we start at 5, move 4 units to the left, and then move 1 more unit to the left.

We arrive at 0. So, $5+(-4)+(-1) = \mathbf{0}$.

— *or* —

$5+(-4)+(-1) = 5+(-4+(-1)) = 5+(-5) = \mathbf{0}$.

66. To add $-3+(-6)+3$, we start at -3, move 6 units to the left, and then move 3 units to the right.

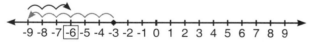

We arrive at -6. So, $-3+(-6)+3 = \mathbf{-6}$.

— *or* —

$(-3)+(-6)+3 = -3+3+(-6)$.
Since $-3+3 = 0$, we have $-3+(-6)+3 = (-3+3)+(-6)$
$= 0+(-6)$
$= \mathbf{-6}$.

Integer-Tac-Toe 82-83

67. Winnie will win if she places a 2 as shown below. Then, the sum of the winning line is $2+(-1)+(-1) = 0$.

68. $-1+3+(-2) = 0$.

69. $-2+1+\mathbf{1} = 0$.

70. $2+(-3)+1 = 0$.

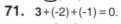

71. $3+(-2)+(-1) = 0$.

72. $1+(-3)+2 = 0$.

73. Grogg will win if he places a -1 as shown below. Then, the sum of the winning line is $2+(\mathbf{-1})+(-1) = 0$.

74. $-1+(-1)+2 = 0$.

75. $3+(-1)+(\mathbf{-2}) = 0$.

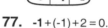

76. $-1+3+(\mathbf{-2}) = 0$.

77. $\mathbf{-1}+(-1)+2 = 0$.

78. $-1+3+(\mathbf{-2}) = 0$.

79. We notice that no winning line can have three negative integers or three positive integers.

So, we look at the possibilities when the winning line has one positive integer and two negative integers.

There are no winning lines that have 1 as the only positive integer, since no two negative integers have a sum of -1.

If 2 is the only positive integer, then the two negative integers must be -1 and -1.

If 3 is the only positive integer, then the two negative integers must be -1 and -2.

Then, we look at the possibilities when the winning line has one negative integer and two positive integers.

There are no winning lines that have -1 as the only negative integer, since no two positive integers have a sum of 1.

If -2 is the only negative integer, then the two positive integers must be 1 and 1.

If -3 is the only negative integer, then the two positive integers must be 1 and 2.

There are no other possible winning lines.

All together, we have **4** different groups of three integers that create a winning line in Integer-Tac-Toe:
(2, -1, -1), (3, -1, -2), (-2, 1, 1), and (-3, 1, 2).

Notice that after we found the winning lines with one positive, we could make the winning lines with one negative by switching each number in the group to its opposite: negative numbers become positive and positive become negative.

80. Winnie cannot win during this turn. However, we notice that if Grogg is able to place a -2 in the upper-left corner, then he will win on his next turn:

-2	3	-1
	1	
	-1	

Winnie can prevent Grogg from winning on his next turn by blocking this spot. She can place a 1, 2, or 3 in this spot, as shown below.

1	3	-1		2	3	-1		3	3	-1
	1		or		1		or		1	
	-1				-1				-1	

However, if Winnie places a 1 or 2 in that spot, we see that Grogg could still win on his next turn.

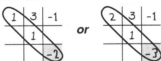

Therefore, Winnie must place a 3 in the upper-left corner to prevent Grogg from winning on his next turn.

3	3	-1
	1	
	-1	

81. Grogg can win on his next turn if he sets up *two or more* possible winning lines but Winnie can only block one.

We see that if Grogg places a -1 in the bottom-left corner, as shown, then he has two possible winning lines: the left column or the bottom-left-to-top-right diagonal.

Winnie does not have a winning move on this board. During her next move, she can block one but not both of Grogg's possible winning lines.

Therefore, no matter what Winnie's next move is, Grogg can guarantee a win by placing a -1 in the bottom-left corner.

None of Grogg's other possible moves guarantee a win.

82. We start at 19 and move 11 units to the left. This brings us 11 units closer to zero. We end up 19 − 11 = 8 units to the right of zero at **8**.

83. We start at -11 and move 17 units to the left. This brings us 17 units farther from zero. We end up 11 + 17 = 28 units to the left of zero at **-28**.

84. We start at -20 and move 13 units to the left. This brings us 13 units farther from zero. We end up 20 + 13 = 33 units to the left of zero at **-33**.

85. We start at -23 and move 35 units to the right. We first move 23 units to the right to arrive at 0, and then we move 35 − 23 = 12 more units to the right. We end up 12 units to the right of zero at **12**.

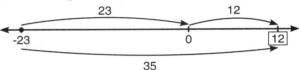

86. We start at -12 and move 71 units to the left. This brings us 71 units farther from zero. We end up 12 + 71 = 83 units to the left of zero at **-83**.

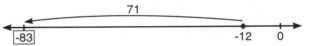

87. We start at 28 and move 14 units to the left. This brings us 14 units closer to zero. We end up 28 − 14 = 14 units to the right of zero at **14**.

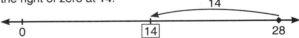

88. We start at 25 and move 16 units to the right. This brings us 16 units farther from zero. We end up 25 + 16 = 41 units to the right of zero at **41**.

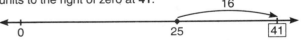

89. We start at 13 and move 48 units to the left. We first move 13 units to the left to arrive at 0, and then we move 48 − 13 = 35 more units to the left. We end up 35 units to the left of zero at **-35**.

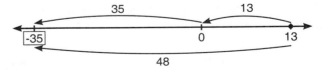

90. When we add any number to its opposite, the result is always zero: -18+18 = **0**.

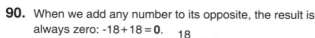

91. We start at 24 and move 36 units to the left. We first move 24 units to the left to arrive at 0, and then we move 36−24 = 12 more units to the left. We end up 12 units to the left of zero, at **-12**.

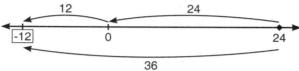

92. The number that is 35 more than -14 is -14+35 = **21**.

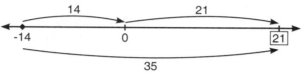

93. The sum of -11 and -19 is -11+(-19) = -30.

When we add 20 to this sum, the result is -30+20 = **-10**.

94. We consider -16,987+6,654. If we start 16,987 units to the left of 0 and only move 6,654 units to the right, then we do not reach or pass 0.

So, -16,987+6,654 is negative.
Similarly, if we start with any negative number and add a positive number which is closer to zero, the result is negative.

Next, we consider -42,345+(-57,654). When we start at a negative number and add another negative number, the result is negative.

So, -42,345+(-57,654) is negative.

Then, we consider 856,915+(-532,812). If we start 856,915 units to the right of 0 and only move 532,812 units to the left, then we do not reach or pass 0.

So, 856,915+(-532,812) is positive.
Similarly, if we start with any positive number and add a negative number which is closer to zero, the result is positive.

Finally, we consider 1,098,765+(-2,345,678). If we start 1,098,765 units to the right of zero and move 2,345,678 units to the left, then we cross zero and continue moving left.

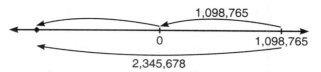

So, 1,098,765+(-2,345,678) is negative.
Similarly, if we start with any positive number and add a negative number which is further from zero, the result is negative.

We circle the sums that are negative:

(-16,987+6,654)   (-42,345+(-57,654))
856,915+(-532,812)   (1,098,765+(-2,345,678))

95. Each term in the pattern is 5 more than the one before it.

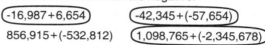

So, we continue the pattern by adding 5's.

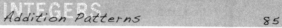

96. We continue the pattern by adding 10's.

+10 +10 +10 +10 +10 +10 +10 +10
-53, -43, -33, **-23**, **-13**, **-3**, **7**, **17**, **27**

97. We continue the pattern by adding 13's.

+13 +13 +13 +13 +13 +13 +13 +13
-55, -42, -29, **-16**, **-3**, **10**, **23**, **36**, **49**

98. To get from -12 to -10, we add 2.

To confirm, we check that to get from -16 to -12, we can add 2 twice: -16+2+2 = -16+4 = -12.

So, each term in the pattern is 2 more than the one before it. We continue the pattern by adding 2's.

+2 +2 +2 +2 +2 +2 +2 +2
-16, **-14**, -12, **-10**, **-8**, **-6**, **-4**, **-2**, **0**

99. To get from -10 to -4, we add 6. To add 6, we can add 3 twice.

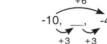

To confirm, we check that to get from -19 to -10, we can add 3 three times: -19+3+3+3 = -19+9 = -10.

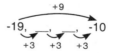

So, each term in this pattern is 3 more than the one before it. We continue the pattern by adding 3's.

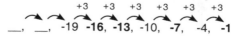

To find the term that comes after -19, we add 3. Then, to find the term that comes *before* -19, we can add -3.

Similarly, we can find the first number in the pattern by adding -3 as we move backwards.

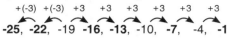

100. To get from -24 to -14, we add 10. To add 10, we can add 5 twice.

To confirm, we check that to get from -14 to 6, we can add 5 four times: -14+5+5+5+5 = -14+20 = 6.

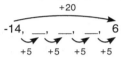

So, each term in the pattern is 5 more than the one before it. We continue the pattern by adding 5's.

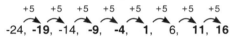

101. To get from -43 to -11, we add 32. To add 32, we can add 8 four times.

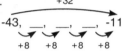

To confirm, we check that to get from -11 to 21, we can add 8 four times: -11+8+8+8+8 = -11+32 = 21.

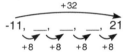

So, each term in the pattern is 8 more than the one before it. We continue the pattern by adding 8's.

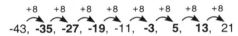

INTEGERS
Sum Squares 86-89

102. First, we look at the left column. We have 1+6+☐ = 16. This simplifies to 7+☐ = 16. Since 7+**9** = 16, we fill in the bottom-left square with 9 as shown.

	16	9	20
12	1		
13	6	2	
20	**9**		7

We look at the bottom row. We have 9+☐+7 = 20. This simplifies to 16+☐ = 20. Since 16+**4** = 20, we fill in the bottom-middle square with 4 as shown.

	16	9	20
12	1		
13	6	2	
20	9	**4**	7

We look at the middle row. We have 6+2+☐ = 13. This simplifies to 8+☐ = 13. Since 8+**5** = 13, we fill in the middle-right square with 5 as shown.

	16	9	20
12	1		
13	6	2	**5**
20	9	4	7

Then, in the middle column, we have **3**+2+4 = 9. In the right column, we have **8**+5+7 = 20.

All digits have been placed. We check the sums of the integers in each row and column.

	16	9	20
12	1	**3**	**8**
13	6	2	5
20	9	4	7

103. First, we look at the middle column. We have -6+4+☐ = -7. This simplifies to -2+☐ = -7. Since -2+**-5** = -7, we fill in the bottom-middle square as shown.

	3	-7	15
9	8	-6	
0		4	
2		**-5**	9

Next, we look at the top row. We have 8+(-6)+☐ = 9, which simplifies to 2+☐ = 9. Since 2+**7** = 9, we fill in the top-right square with 7 as shown.

	3	-7	15
9	8	-6	**7**
0		4	
2		-5	9

Then, we look at the bottom row. We have ☐+(-5)+9 = 2, which simplifies to ☐+4 = 2. Since **-2**+4 = 2, we fill in the bottom-left square with -2 as shown.

	3	-7	15
9	8	-6	7
0		4	
2	**-2**	-5	9

Then, in the left column, we have 8+**-3**+(-2) = 3. In the right column, we have 7+**-1**+9 = 15.

All digits have been placed. We check the sums of the integers in each row and column.

	3	-7	15
9	8	-6	7
0	**-3**	4	**-1**
2	-2	-5	9

We solve the Sum Squares below with the steps shown, using the same strategies as in the previous problems.

104. Step 1:

	9	0	0
9	6	5	**-2**
0		**3**	-7
0			-8

Step 2:

	9	0	0
9	6	5	-2
0	**4**	3	-7
0		-8	**9**

Final:

	9	0	0
9	6	5	-2
0	4	3	-7
0	**-1**	-8	9

105. Step 1:

	-8	-10	-5
-4		8	
-11		1	**-9**
-8	2	**-6**	-4

Step 2:

	-8	-10	-5
-4		**-5**	8
-11	**-3**	1	-9
-8	2	-6	-4

Final:

	-8	-10	-5
-4	**-7**	-5	8
-11	-3	1	-9
-8	2	-6	-4

106. Step 1:

	0	0	3
0	**1**	**3**	-4
0	**7**	-9	
3	-8		

Step 2:

	0	0	3
0	1	3	-4
0	7	-9	**2**
3	-8	**6**	

Final:

	0	0	3
0	1	3	-4
0	7	-9	2
3	-8	6	**5**

107. Step 1:

	20	0	-11
12	9	**7**	-4
0		-2	**-6**
-3			-1

Step 2:

	20	0	-11
12	9	7	-4
0	**8**	-2	-6
-3		**-5**	-1

Final:

	20	0	-11
12	9	7	-4
0	8	-2	-6
-3	**3**	-5	-1

108. In the top row, we have $\boxed{9}+(-3)+(-5)=1$.
The remaining digits are 1, 4, 6, and 7.

	-1	2	8
1	9	-3	-5
3	-8		
5	-2		

The two missing entries in the middle row must sum to 11 because $-8+\boxed{11}=3$. The only way to make a sum of 11 from two of the remaining the digits is $4+7=11$.

	-1	2	8
1	9	-3	-5
3	-8		
5	-2		

Similarly, the two missing entries in the right column must sum to 13 because $-5+\boxed{13}=8$. The only way to make a sum of 13 from two of the remaining digits is $6+7=13$.

	-1	2	8
1	9	-3	-5
3	-8		7
5	-2		6

We learned above that 7 is in the middle row. So, we place the 7 as shown in the middle row, right column, with the 6 below it.

In the middle row, we have $-8+\boxed{4}+7=3$.
In the bottom row, we have $-2+\boxed{1}+6=5$.
All digits have been placed. We check the sums of the integers in each row and column.

	-1	2	8
1	9	-3	-5
3	-8	4	7
5	-2	1	6

109. Step 1:

	-4	10	-11
4	-7	8	
-12		-4	
3		6	

Step 2:

	-4	10	-11
4	-7	8	3
-12		-4	
3		6	

Step 3:
The remaining digits are 1, 2, 5, and 9.

The two missing entries in the right column sum to -14 because $3+\boxed{-14}=-11$. The only way to make a sum of -14 from two of the remaining digits is $-5+(-9)=-14$.

	-4	10	-11
4	-7	8	3
-12		-4	
3		6	

Similarly, the two missing entries in the middle row sum to -8 because $-4+\boxed{-8}=-12$. The only way to make a sum of -8 from two of the remaining digits is $-9+1=-8$.

	-4	10	-11
4	-7	8	3
-12		-4	-9
3		6	

We learned above that -9 is in the right column. So, we place the -9 as shown.

Then, we fill in the remaining entries as shown.

Step 4:

	-4	10	-11
4	-7	8	3
-12	1	-4	-9
3		6	-5

Final:

	-4	10	-11
4	-7	8	3
-12	1	-4	-9
3	2	6	-5

110. Step 1:

	7	7	7
1	9	-7	-1
11	2		
9	-4		

Step 2:
The remaining digits are 3, 5, 6, and 8.

The two missing entries in the center column sum to 14 because $-7+\boxed{14}=7$. The only way to make a sum of 14 from two of the remaining digits is $6+8=14$.

	7	7	7
1	9	-7	-1
11	2		
9	-4		

Similarly, the two missing entries in the middle row sum to 9 because $2+\boxed{9}=11$. The only way to make a sum of 9 from two of the remaining digits is $3+6=9$.

	7	7	7
1	9	-7	-1
11	2	6	
9	-4		

We learned above that 6 is in the middle column. So, we place the 6 as shown.

Then, we fill in the remaining entries as shown.

Step 3:

	7	7	7
1	9	-7	-1
11	2	6	3
9	-4	8	

Final:

	7	7	7
1	9	-7	-1
11	2	6	3
9	-4	8	5

111. Step 1:

	8	-5	-2
-10	5	-8	-7
10			
1		4	

Step 2:

	8	-5	-2
-10	5	-8	-7
10		-1	
1		4	

Step 3:
The remaining digits are 2, 3, 6, and 9.

The two missing entries in the right column sum to 5 because $-7+\boxed{5}=-2$. The only way to make a sum of 5 from two of the remaining digits is $2+3=5$.

	8	-5	-2
-10	5	-8	-7
10		-1	
1		4	

Similarly, the two missing entries in the middle row sum to 11 because $-1+\boxed{11}=10$. The only way to make a sum of 11 from two of the remaining digits is $2+9=11$.

	8	-5	-2
-10	5	-8	-7
10		-1	2
1		4	

We learned above that 2 is in the right column. So, we place the 2 as shown.

Then, we fill in the remaining entries as shown.

Step 4:

	8	-5	-2
-10	5	-8	-7
10	9	-1	2
1		4	3

Final:

	8	-5	-2
-10	5	-8	-7
10	9	-1	2
1	-6	4	3

112. Step 1:

	-2	5	6
-1	1	2	-4
2			
8	-8		

Step 2:

	-2	5	6
-1	1	2	-4
2		5	
8	-8		

Step 3:

The remaining digits are 3, 6, 7, and 9.

The two missing entries in the right column sum to 10 because -4+$\boxed{10}$=6. The only way to make a sum of 10 from two of the remaining digits is 7+3=10.

Similarly, the two missing entries in the bottom row sum to 16 because -8+$\boxed{16}$=8. The only way to make a sum of 16 from two of the remaining digits is 7+9=16.

We learned above that 7 is in the right column. So, we place the 7 as shown.

Then, we fill in the remaining entries as shown.

Step 4:   Final:

113. Step 1:   Step 2:

Step 3:

The remaining digits are 2, 3, 6, and 7.

The two missing entries in the right column sum to 5 because -1+$\boxed{5}$=4. We can make a sum of 5 from two of the remaining digits in two ways: 2+3=5 or -2+7=5.

So, the digit 6 *will not* be in the right column.

The two missing entries in the middle row sum to -5 because -4+$\boxed{-5}$=-9. We can make a sum of -5 from two of the remaining digits in two ways: -2+(-3)=-5 or 2+(-7)=-5.

So, the digit 6 *will not* be in the middle row.

We learned above that the digit 6 will not be in the right column or the middle row. Therefore, the top-left corner contains 6 or -6.

The two missing entries in the left column sum to -9 because 9+$\boxed{-9}$=0. The only way to make a sum of -9 from the digit 6 and one of the remaining digits is -6+(-3)=-9.

So, we place the -6 as shown.

Then, we fill in the remaining entries as shown.

Step 4:   Final:

114. Step 1:

The two missing entries in the top row sum to 16 because -5+$\boxed{16}$=11. The only way to make a sum of 16 from two of the remaining digits is 7+9=16.

Similarly, the two missing entries in the right column sum to 17 because -3+$\boxed{17}$=14. The only way to make a sum of 17 from two of the remaining digits is 8+9=17.

We learned above that 9 is in the top row. So, we place the 9 as shown.

Then, we fill in the remaining entries as shown.

Step 2:   Step 3:   Final:

115. Step 1:

The remaining digits are 2, 3, 4, 6, 7, and 8.

The two missing entries in the middle column sum to 6 because 1+$\boxed{6}$=7. We can make a sum of 6 from two of the remaining digits in two ways: 2+4=6 or -2+8=6.

The two missing entries in the top row sum to 9 because -5+$\boxed{9}$=4. We can make a sum of 9 from two of the remaining digits in two ways: 2+7=9 or 3+6=9.

We learned above that only 2, 4, -2, or 8 could be placed in the middle column.

So, the only number that can be placed in the top row and in the middle column is 2. We place it as shown.

Then, we fill in the remaining entries as shown.

Step 2:   Step 3:   Final:

116. 7+8+9=**24** is the greatest possible sum of the numbers in a row or column of a Sum Square.

117. $-7+(-8)+(-9) = $ **-24** is the least possible sum of the numbers in a row or column of a Sum Square.

118. The sum of the three numbers in the middle row is -23. The only way we can get this sum with three numbers in our puzzle is $-6+(-8)+(-9) = -23$.

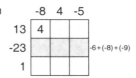

So, the middle row contains -6, -8, and -9, in some order.

The remaining digits are 1, 2, 3, 5, and 7.

The two missing entries in the top row sum to 9 because $4+\boxed{9} = 13$. The only way to make a sum of 9 from two of the remaining digits is $2+7=9$.

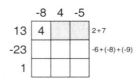

So, the top row contains 2 and 7, in some order.

The remaining digits are 1, 3, and 5. These digits must appear in the bottom row.

The only way to make a sum of 1 from these digits is $-1+(-3)+5=1$.

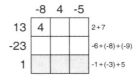

So, the bottom row contains -1, -3, and 5, in some order.

The two missing entries in the left column sum to -12 because $4+\boxed{-12} = -8$. The only way to make a sum of -12 from a middle-row digit and a bottom-row digit is $-9+(-3) = -12$.

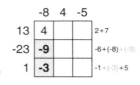

So, we place the digits as shown.

We look at the middle column.

If -1 is in the bottom square of the middle column, then the two missing entries in the middle column sum to 5 because $-1+\boxed{5} = 4$.

We cannot make a sum of 5 from one number in the top row and one number in the middle row. So, -1 cannot be in the middle column.

Therefore, 5 is in the bottom row, middle column, and -1 is in the bottom row, right column.

The two missing entries in the middle column sum to -1 because $5+\boxed{-1} = 4$. The only way to make a sum of -1 from two of the remaining digits is $7+(-8) = -1$.

We know 7 is in the top row and -8 is in the middle row, so we place these numbers as shown.

Finally, we know that 2 is the missing entry in the top row, and -6 is the missing entry in the middle row.

We check that $2+(-6)+(-1) = -5$, and we are done.

119. The sum of the numbers in the right column is 24. The only way we can get this sum with three numbers in our puzzle is $7+8+9 = 24$.

So, the right column contains 7, 8, and 9, in some order. The remaining digits are 1, 2, 3, 4, 5, and 6.

We look at the top row.

If 9 is in the top row, right column, then the two missing entries in the top row sum to -13 because $9+\boxed{-13} = -4$.

We cannot make a sum of -13 from any two of the remaining digits. So, 9 cannot be in the top row.

Similarly, if 8 is in the top row, right column, then the two missing entries in the top row sum to -12 because $8+\boxed{-12} = -4$.

We cannot make a sum of -12 from any two of the remaining digits. So, 8 cannot be in the top row.

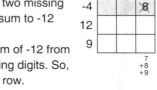

Therefore, 7 is in the top square of the right column.

We look at the bottom row.

If 9 is in the bottom square of the right column, then the two missing entries in the bottom row sum to 0 because $9+\boxed{0} = 9$.

We cannot make a sum of 0 from any two different digits. So, 9 cannot be in the bottom row.

Therefore, 8 is in the bottom square of the right column, and 9 is in the middle square of the right column.

The two missing entries in the top row sum to -11 because $7+\boxed{-11} = -4$. The only way to make a sum of -11 from two of the remaining digits is $-5+(-6) = -11$.

So, the top row contains -5 and -6 in some order. The remaining digits are 1, 2, 3, and 4.

We look at the middle column.

If -6 is in the top square of the middle column, then the two missing entries in the middle column sum to 8 because -6+$\boxed{8}$=2.

We cannot make a sum of 8 from any two of the remaining digits (1, 2, 3, or 4). So, -6 cannot be in the middle column.

Therefore, we place the -6 and -5 as shown.

The two missing entries in the middle column sum to 7 because -5+$\boxed{7}$=2. The only way we can make a sum of 7 from two of the remaining digits is 3+4=7.

If 3 is in the center, then the remaining entry in the middle row is 0 because 3+9+$\boxed{0}$=12.

However, 0 is not a digit we can place in this puzzle. So, 3 cannot be in the center.

Therefore, we place the 4 in the center of the square, with the 3 below it as shown.

Then, we fill in the remaining entries as shown.

Subtracting Integers 90-93

120. To subtract 2−7, we start at 2 and move 7 units to the left. We arrive at -5. So, 2−7 = **-5**.

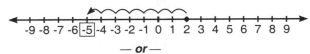

— or —

We start at 2 and move 7 units to the left. We first move 2 units to the left to arrive at 0, and then we move 7−2 = 5 more units to the left. We end up 5 units to the left of zero at **-5**.

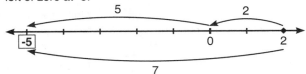

121. To subtract -2−6, we start at -2 and move 6 units to the left. We arrive at -8, so -2−6 = **-8**.

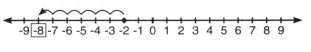

— or —

We start at -2 and move 6 units to the left. This brings us 6 units farther from zero. We end up 2+6 = 8 units to the left of zero, at **-8**.

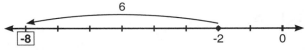

We use the approaches described in the previous solutions to compute the following differences.

122. -4−3 = **-7**.　　**123.** 9−14 = **-5**.
124. 5−11 = **-6**.　　**125.** 8−1 = **7**.
126. 6−10 = **-4**.　　**127.** -3−6 = **-9**.
128. 0−4 = **-4**.　　**129.** -1−1 = **-2**.

130. To subtract an integer, we add its opposite. So, to subtract -2, we add 2. Therefore, we have
$$9-(-2) = \underline{9}+\underline{2} = \underline{\mathbf{11}}.$$

131. To subtract -3, we add 3. Therefore, we have
$$-6-(-3) = \underline{-6}+\underline{3} = \underline{\mathbf{-3}}.$$

132. To subtract -9, we add 9. Therefore, we have
$$6-(-9) = \underline{6}+\underline{9} = \underline{\mathbf{15}}.$$

133. To subtract 11, we add -11. Therefore, we have
$$-5-11 = \underline{-5}+\underline{(-11)} = \underline{\mathbf{-16}}.$$

134. To subtract -1, we add 1. Therefore, we have
$$-8-(-1) = \underline{-8}+\underline{1} = \underline{\mathbf{-7}}.$$

135. 12−17 = **-5**.

136. -11−15 = **-26**.

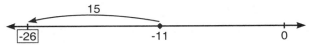

137. 18−(-6) = 18+6 = **24**.

138. -9−(-14) = -9+14 = **5**.

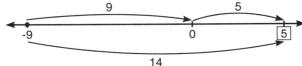

139. -17−(-7) = -17+7 = **-10**.

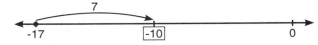

140. 18 − 30 = **-12**.

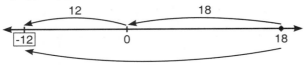

141. 0 − (-11) = 0 + 11 = **11**.

142. -13 − 13 = **-26**.

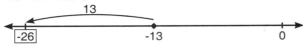

143. -22 − (-14) = -22 + 14 = **-8**.

144. -4 − (-4) = -4 + 4 = **0**.

— or —

The result of subtracting a number from itself is always 0. So, -4 − (-4) = **0**.

145. -790 − 11 = **-801**.

146. -23 − (-314) = -23 + 314 = **291**.

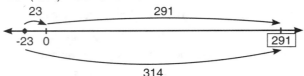

147. Since -4 + 9 = 5, we have
⎣5⎦ − ⎣9⎦ = ⎣-4⎦ and ⎣5⎦ − ⎣-4⎦ = ⎣9⎦.

148. Since -5 + 2 = -3, we have
⎣-3⎦ − ⎣2⎦ = ⎣-5⎦ and ⎣-3⎦ − ⎣-5⎦ = ⎣2⎦.

149. Since -1 + 4 = 3, we have
⎣3⎦ − ⎣4⎦ = ⎣-1⎦ and ⎣3⎦ − ⎣-1⎦ = ⎣4⎦.

150. Since -13 + 8 = -5, we have
⎣-5⎦ − ⎣8⎦ = ⎣-13⎦ and ⎣-5⎦ − ⎣-13⎦ = ⎣8⎦.

151. Since -4 + (-10) = -14, we have
⎣-14⎦ − ⎣-10⎦ = ⎣-4⎦ and ⎣-14⎦ − ⎣-4⎦ = ⎣-10⎦.

152. Since -7 + (-9) = -16, we have
⎣-16⎦ − ⎣-9⎦ = ⎣-7⎦ and ⎣-16⎦ − ⎣-7⎦ = ⎣-9⎦.

INTEGERS — Subtraction Patterns · 94

153. Each term in the pattern is 10 less than the one before it.

       −10 −10
  32, 22, 12, __, __, __, __, __

So, we continue the pattern by subtracting 10's.

       −10 −10 −10 −10 −10 −10 −10 −10
  32, 22, 12, **2**, **-8**, **-18**, **-28**, **-38**, **-48**

154. We continue the pattern by subtracting 5's.

       −5 −5 −5 −5 −5 −5 −5 −5
  17, 12, 7, **2**, **-3**, **-8**, **-13**, **-18**, **-23**

155. We continue the pattern by subtracting 11's.

       −11 −11 −11 −11 −11 −11 −11 −11
  20, 9, -2, **-13**, **-24**, **-35**, **-46**, **-57**, **-68**

156. To get from 3 to -1, we subtract 4.

To confirm, we check that to get from 11 to 3, we can subtract 4 two times.

          −8
  11, __, 3
      −4 −4

So, each term in the pattern is 4 less than the one before it. We continue the pattern by subtracting 4's.

       −4 −4 −4 −4 −4 −4 −4 −4
  11, **7**, 3, -1, **-5**, **-9**, **-13**, **-17**, **-21**

157. To get from -35 to -51, we subtract 16. To subtract 16, we can subtract 8 twice.

To confirm, we check that to get from -11 to -35, we can subtract 8 three times.

          −24
  -11, __, __, -35
       −8 −8 −8

So, each term in this pattern is eight less than the one before it. We continue the pattern by subtracting 8's.

       −8 −8 −8 −8 −8 −8
  __, __, -11, **-19**, **-27**, -35, **-43**, -51, **-59**

To find the term that comes after -11, we subtract 8. To find the term that comes *before* -11, we can add 8.

Similarly, we can find the first number in the pattern by adding 8 as we move backwards.

     +8 +8 −8 −8 −8 −8 −8 −8
  5, **-3**, -11, **-19**, **-27**, **-35**, **-43**, **-51**, **-59**

158. To get from 29 to -4, we subtract 33. To subtract 33, we can subtract 11 three times.

To confirm, we check that to get from -4 to -48, we can subtract 11 four times.

So, each term in the pattern is eleven less than the one before it. We continue the pattern by subtracting 11's.

       −11 −11 −11 −11 −11 −11 −11 −11
  29, **18**, **7**, -4, **-15**, **-26**, **-37**, **-48**, **-59**

159. To get from 13 to -15, we subtract 28. To subtract 28, we can subtract 7 four times.

To confirm, we check that to get from -15 to -43, we can subtract 7 four times.

Therefore, each term in the pattern is seven less than the one before it. We continue the pattern by subtracting 7's.

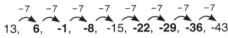

Cross-Number Puzzles 95-97

We show one way to find the missing entries of these puzzles. You may have solved each in a different order.

160. We fill in the missing entries as shown below.

$9+(-2)=\boxed{7}$. $9-(-3)=\boxed{12}$. $7-2=\boxed{5}$, or
$-3+5=\boxed{2}$. $-2-5=\boxed{-7}$. $12+(-7)=\boxed{5}$.

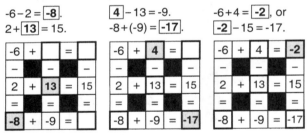

161. We fill in the missing entries as shown below.

$-6-2=\boxed{-8}$. $\boxed{4}-13=-9$. $-6+4=\boxed{-2}$, or
$2+\boxed{13}=15$. $-8+(-9)=\boxed{-17}$. $\boxed{-2}-15=-17$.

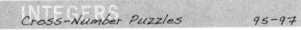

For the remaining puzzles, we show the final answer.

162.–165.

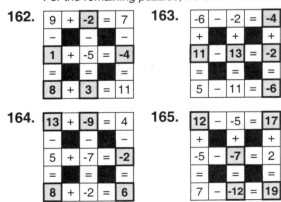

166.–173.

Absolute Value 98-100

174. Since 26 is 26 units from zero, the absolute value of 26 is 26. Therefore, we have |26| = **26**.

175. Since -8 is 8 units from zero, the absolute value of -8 is 8. Therefore, we have |-8| = **8**.

176. Since -35 is 35 units from zero, we have |-35| = **35**.

177. Since 72 is 72 units from zero, we have |72| = **72**.

178. |-3| = 3. Since 3 > -4, we have |-3| **>** -4.

179. |-17| = 17 and |17| = 17. Since 17 = 17, we have |-17| **=** |17|.

180. |-54| = 54, and |-19| = 19. Since 54 > 19, we have |-54| **>** |-19|.

181. |-13| = 13. Since 12 < 13, we have 12 **<** |-13|.

182. On the number line, the integer 15 units to the left of zero is -15, and the integer 15 units to the right of zero is 15. So, **-15 and 15** are the only two integers that have absolute value 15.

183. The integers with absolute value 0 are the integers that are 0 units from zero. The only integer that fits that description is 0. So, there is just **one** integer with absolute value 0.

184. When we measure a number's distance from zero, the distance will always be positive or 0. No number has a negative distance from zero, so no number has a negative absolute value. Therefore, **zero** integers have absolute value -15.

185. First, we evaluate the part of the expression between the absolute value bars: $|\text{-}5-3| = |\text{-}8|$.

Then, the absolute value of -8 is 8. So, we have
$$|\text{-}5-3| = |\text{-}8| = \mathbf{8}.$$

186. First, the absolute value of -5 is 5, so $|\text{-}5|-3 = 5-3$. Then, we subtract. So, we have
$$|\text{-}5|-3 = 5-3 = \mathbf{2}.$$

187. We have $|\text{-}10|-16 = 10-16 = \mathbf{\text{-}6}$.

188. We have $|\text{-}10-16| = |\text{-}26| = \mathbf{26}$.

189. We have $|2-5|-11 = |\text{-}3|-11 = 3-11 = \mathbf{\text{-}8}$.

190. We have $2-|5-11| = 2-|\text{-}6| = 2-6 = \mathbf{\text{-}4}$.

191. We have $|\text{-}13+3-14| = |\text{-}10-14| = |\text{-}24| = \mathbf{24}$.

192. We have $\text{-}13+|3-14| = \text{-}13+|\text{-}11| = \text{-}13+11 = \mathbf{\text{-}2}$.

193. We have $|\text{-}6-3|-4 = |\text{-}9|-4 = 9-4 = \mathbf{5}$.

194. We have $\text{-}6-|3-4| = \text{-}6-|\text{-}1| = \text{-}6-1 = \mathbf{\text{-}7}$.

195. Since $4+\boxed{1}=5$, we know that $|x|=1$. The only two numbers with absolute value 1 are 1 and -1.
So, there are two values of x for which $4+|x|=5$:
$x = \mathbf{1}$ and $x = \mathbf{\text{-}1}$.

196. No number has a negative absolute value. Therefore, there is no number whose absolute value is -4, and no number makes the equation $|x|=\text{-}4$ true. Therefore, the equation is **impossible**.

197. Since $\boxed{9}-5=4$, we know that $|x|=9$. The only two numbers with absolute value 9 are 9 and -9.
So, there are two values of x for which $|x|-5=4$:
$x = \mathbf{9}$ and $x = \mathbf{\text{-}9}$.

198. Since $\boxed{\text{-}1}+5=4$, we know that the values of x that make this equation true are the values of x for which $|x|=\text{-}1$. However, no number has a negative absolute value. So, there is no number whose absolute value is -1. No number makes the equation $|x|+5=4$ true. Therefore, the equation is **impossible**.

199. Since $5-\boxed{1}=4$, we know that $|x|=1$. The only two numbers with absolute value 1 are 1 and -1.
So, there are two values of x for which $5-|x|=4$:
$x = \mathbf{1}$ and $x = \mathbf{\text{-}1}$.

200. Since $4-\boxed{\text{-}1}=5$, we know that the values of x that make this equation true are the values of x for which $|x|=\text{-}1$. However, no number has a negative absolute value. So, there is no number whose absolute value is -1. No number makes the equation $4-|x|=5$ true. Therefore, the equation is **impossible**.

201. The only two numbers with absolute value 5 are 5 and -5. So, $|x-4|=5$ tells us that $x-4$ is equal to 5 or -5.

If $x-4=5$, then $x=9$.
If $x-4=\text{-}5$, then $x=\text{-}1$.

So, there are two values of x for which $|x-4|=5$:
$x = \mathbf{9}$ and $x = \mathbf{\text{-}1}$.

202. The only two numbers with absolute value 5 are 5 and -5. So, $|4-x|=5$ tells us that $4-x$ is equal to 5 or -5.

If $4-x=5$, then $x=\text{-}1$.
If $4-x=\text{-}5$, then $x=9$.

So, there are two values of x for which $|4-x|=5$:
$x = \mathbf{\text{-}1}$ and $x = \mathbf{9}$.

We notice that the solutions to $|4-x|=5$ and $|x-4|=5$ (from the previous problem) are the same!

Adding to Subtract

203. We rewrite this problem as a sum:
$$183+120-153 = 183+120+(\text{-}153).$$

In our sum, there are two terms that end in 3: one positive and one negative.

$$183+120+(\text{-}153)$$

Since these two terms will be easiest to add, we rearrange to pair these terms. Then, we evaluate:
$$\begin{aligned}183+120-153 &= 183+120+(\text{-}153)\\&= 183+(\text{-}153)+120\\&= 30+120\\&= \mathbf{150}.\end{aligned}$$

204. We rewrite this problem as a sum:
$$159+217-49 = 159+217+(\text{-}49).$$

In our sum, there are two terms that end in 9: one positive and one negative.

$$159+217+(\text{-}49)$$

Since these two terms will be easiest to add, we rearrange to pair these terms. Then, we evaluate:
$$\begin{aligned}159+217-49 &= 159+217+(\text{-}49)\\&= 159+(\text{-}49)+217\\&= 110+217\\&= \mathbf{327}.\end{aligned}$$

205. We rewrite this problem as a sum:
$$10+20+30-10-20 = 10+20+30+(\text{-}10)+(\text{-}20).$$

In our sum, there are two pairs of numbers that are opposites.

$$10+20+30+(\text{-}10)+(\text{-}20)$$

Since the sum of each pair is 0, we rearrange to pair these terms. Then, we evaluate:
$$\begin{aligned}10+20+30-10-20 &= 10+20+30+(\text{-}10)+(\text{-}20)\\&= 10+(\text{-}10)+20+(\text{-}20)+30\\&= 0+0+30\\&= \mathbf{30}.\end{aligned}$$

206. We rewrite this problem as a sum:
$$7-6+8-7+9-8 = 7+(\text{-}6)+8+(\text{-}7)+9+(\text{-}8).$$

In our sum, there are three pairs of numbers whose sum is 1.

$$7+(\text{-}6)+8+(\text{-}7)+9+(\text{-}8)$$

Since these pairs will be easiest to add, we group them:
$$\begin{aligned}7-6+8-7+9-8 &= 7+(\text{-}6)+8+(\text{-}7)+9+(\text{-}8)\\&= 1+1+1\\&= \mathbf{3}.\end{aligned}$$

— or —

We rewrite this problem as a sum:

$7-6+8-7+9-8 = 7+(-6)+8+(-7)+9+(-8)$.

In our sum, there are two pairs of numbers that are opposites.

$7+(-6)+8+(-7)+9+(-8)$

Since the sum of each pair is 0, we rearrange to pair these terms. Then, we evaluate:

$7-6+8-7+9-8 = 7+(-7)+8+(-8)+(-6)+9$
$= 0 + 0 +(-6)+9$
$= 3$.

207. We rewrite this problem as a sum:

$5+15+25+35-10-20-30$
$= 5+15+25+35+(-10)+(-20)+(-30)$.

In our sum, there are three pairs of numbers whose sum is 5.

$5+15+25+35+(-10)+(-20)+(-30)$

Since these pairs will be easiest to add, we rearrange to group them. Then, we evaluate:

$5+15+25+35-10-20-30$
$= 5+15+25+35+(-10)+(-20)+(-30)$
$= 5+15+(-10)+25+(-20)+35+(-30)$
$= 5+ 5 + 5 + 5$
$= 20$.

208. We rewrite this problem as a sum:

$28+48+68+88-18-28-38-48$
$= 28+48+68+88+(-18)+(-28)+(-38)+(-48)$.

In our sum, there are two pairs of numbers that are opposites.

$28+48+68+88+(-18)+(-28)+(-38)+(-48)$

Since the sum of each pair is 0, we rearrange to group them. We also have two pairs of numbers whose sum is 50.

$28+(-28)+48+(-48)+68+88+(-18)+(-38)$

So, we rearrange to group these pairs. Then, we evaluate:

$28+48+68+88-18-28-38-48$
$= 28+48+68+88+(-18)+(-28)+(-38)+(-48)$
$= 28+(-28)+48+(-48)+68+(-18)+88+(-38)$
$= 0 + 0 + 50 + 50$
$= 100$.

— or —

We rewrite this problem as a sum:

$28+48+68+88-18-28-38-48$
$= 28+48+68+88+(-18)+(-28)+(-38)+(-48)$.

All numbers in our sum end in 8, with four positive and four negative. So, we can group four pairs of numbers whose sum is a multiple of 10:

$28+48+68+88+(-18)+(-28)+(-38)+(-48)$

Since these pairs will be easiest to add, we rearrange to group them. Then, we evaluate:

$28+48+68+88-18-28-38-48$
$= 28+48+68+88+(-18)+(-28)+(-38)+(-48)$
$= 28+(-18)+48+(-28)+68+(-38)+88+(-48)$
$= 10 + 20 + 30 + 40$
$= 100$.

209. We rewrite this problem as a sum:

$-418+1,545+2,713-545-1,713+918$
$= -418+1,545+2,713+(-545)+(-1,713)+918$.

In our sum, there is one pair of terms that ends in 18, one pair that ends in 545, and one pair that ends in 713. In each of these pairs, there is one positive and one negative term.

$-418+1,545+2,713+(-545)+(-1,713)+918$

Since these pairs will be easiest to add, we rearrange to group these pairs. Then, we evaluate:

$-418+1,545+2,713-545-1,713+918$
$= -418+1,545+2,713+(-545)+(-1,713)+918$
$= -418+918+1,545+(-545)+2,713+(-1,713)$
$= 500 + 1,000 + 1,000$
$= 2,500$.

210. The negative integers that are greater than -10 are -9, -8, -7, -6, -5, -4, -3, -2, and -1. Their sum is
$-9+(-8)+(-7)+(-6)+(-5)+(-4)+(-3)+(-2)+(-1)$.

We notice there are four pairs of numbers whose sum is -10:

$-9+(-8)+(-7)+(-6)+(-5)+(-4)+(-3)+(-2)+(-1)$

So, we reorder and regroup the addition to make this sum easier to compute:

$-9+(-8)+(-7)+(-6)+(-5)+(-4)+(-3)+(-2)+(-1)$
$= (-9)+(-1)+(-8)+(-2)+(-7)+(-3)+(-6)+(-4)+(-5)$
$= (-10) + (-10) + (-10) + (-10) +(-5)$
$= -45$.

211. The integers greater than -4 but less than 4 are -3, -2, -1, 0, 1, 2, and 3. Their sum is $-3+(-2)+(-1)+0+1+2+3$. We have three pairs of numbers that are opposites: (-3 and 3), (-2 and 2), and (-1 and 1). Since the sum of each pair is 0, we rearrange our sum to pair these terms. Then, we evaluate:
$-3+(-2)+(-1)+0+1+2+3$
$= -3+3+(-2)+2+(-1)+1+0$
$= 0 + 0 + 0 +0$
$= 0$.

212. The distance between -13 and 27 on the number line is $13+27 = 40$ units.

The number that is equal distance from both -13 and 27 lies halfway between these two numbers. Therefore, the number we seek is $40 \div 2 = 20$ units greater than -13 and 20 units less than 27.

We can add $-13+20 = 7$ or subtract $27-20 = 7$. So, **7** is the number that is the same distance from -13 as it is from 27.

213. Alex can group these numbers so that the two groups have an equal sum, so the sum of all 7 numbers is *twice* the sum of each group.

$-9+(-3)+11+5+(-4)+(-10)+14 = 4.$

The total sum of all seven numbers is 4, and we are told that Alex can organize the integers into two groups, each with an equal sum. So, the sum of the numbers in one of the groups is $4 \div 2 = \mathbf{2}$.

We check that such a grouping is possible. If we group (-9, -3, 14) and (11, 5, -4, -10), then we have $-9+(-3)+14 = 2$ and $11+5+(-4)+(-10) = 2$. ✓

214. Since we wish to find the smallest of the five integers, we use a variable to represent that integer: x.

Then, we can write expressions for the next four consecutive integers: $x+1$, $x+2$, $x+3$, and $x+4$.

So, the sum of our five consecutive integers is

$x+(x+1)+(x+2)+(x+3)+(x+4) = -5.$

We simplify the left side to get

$x+x+x+x+x+(1+2+3+4) = -5$
$x+x+x+x+x+10 = -5.$

Subtracting 10 from both sides of the equation and simplifying, we have

$x+x+x+x+x+10-10 = -5-10.$
$x+x+x+x+x = -15.$

We can add five -3's to get -15:
$\boxed{-3}+\boxed{-3}+\boxed{-3}+\boxed{-3}+\boxed{-3} = -15.$

So, $x = -3$. Since x represents the smallest of the five integers, our answer is **-3**.

— **or** —

We define m as the middle number of our five consecutive numbers. So, our five numbers are

$m-2, \ m-1, \ m, \ m+1, \ m+2.$

The sum of these five numbers is
$(m-2)+(m-1)+m+(m+1)+(m+2) = -5.$

We rewrite the expression on the left as a sum:
$m+(-2)+m+(-1)+m+m+1+m+2 = -5.$

We have two pairs of numbers that are opposites: (-2 and 2) and (-1 and 1). Since the sum of each pair is 0, we rearrange our sum to pair these terms.

$m+(-2)+m+(-1)+m+m+1+m+2 = -5.$
$(-2)+2+(-1)+1+m+m+m+m+m = -5.$
$\underbrace{0}_{} + \underbrace{0}_{} +m+m+m+m+m = -5.$

We can add five -1's to get -5:
$\boxed{-1}+\boxed{-1}+\boxed{-1}+\boxed{-1}+\boxed{-1} = -5.$

So, $m = -1$, and our five integers are

$m-2$	$m-1$	m	$m+1$	$m+2$
-3	-2	-1	0	1

The smallest of the five integers is $m-2 = \mathbf{-3}$.

215. The sum of the numbers in each line of three numbers in a row, column, or diagonal is 3. So, we consider the three-number groups of integers whose sum is 3. We organize our work by ordering the numbers in the sum from least to greatest.

For example, if the smallest number is -3, then we know $-3+\boxed{6} = 3$. So, the two other numbers in a row or column containing -3 must sum to 6. There are two groups with smallest integer -3 that sum to 3: (-3, 1, 5) and (-3, 2, 4).

Continuing this strategy, we have 8 possible groups:

(-3, 1, 5), (-3, 2, 4), (-2, 0, 5), (-2, 1, 4), (-2, 2, 3), (-1, 0, 4), (-1, 1, 3), and (0, 1, 2).

We see that the number in the center of the Magic Square is part of four different sums.

Of our available numbers, only 1 appears in at least four of the possible sums. So, we place 1 in the center of the square as shown.

We also see that each number on a corner of the Magic Square is part of three different sums.

Of the remaining numbers, there are exactly four that appear in at least three sums: -2, 0, 2, and 4. So, these four numbers must appear on the four corners of the square.

We choose one of these four numbers to place in the top-left corner. Here, we choose -2.

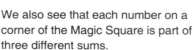

$-2+1+\boxed{4} = 3$. So, to complete the diagonal, the number in the lower-right corner is 4, another one of our corner numbers.

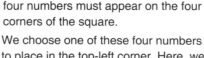

We choose one of the two remaining corner numbers (0 or 2) to place in another corner. Here, we choose 0 and place it in the lower-left corner.

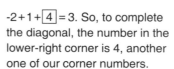

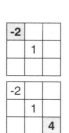

We use the sums of the rows, columns, and diagonals to find the placements of the remaining numbers.

$-2+\boxed{5}+0=3.$

$0+1+\boxed{2}=3.$

$0+\boxed{-1}+4=3.$

-2		2
5	1	
0	-1	4

$-2+\boxed{3}+2=3.$

$5+1+\boxed{-3}=3.$

-2	3	2
5	1	-3
0	-1	4

We could have placed any of the four corner numbers in the top-left corner, and then either of the two remaining corner numbers in the lower-left corner.

So, there are 8 valid arrangements of these numbers into a Magic Square, all shown below. The seven other arrangements are just flipped or rotated versions of our Magic Square above.

-2	3	2		-2	5	0		0	-1	4		0	5	-2
5	1	-3		3	1	-1		5	1	-3		-1	1	3
0	-1	4		2	-3	4		-2	3	2		4	-3	2

2	-3	4		2	3	-2		4	-3	2		4	-1	0
3	1	-1		-3	1	5		-1	1	3		-3	1	5
-2	5	0		4	-1	0		0	5	-2		2	3	-2

— or —

Consider the sum of the numbers in the 9 shaded squares below:

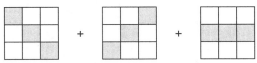

Since the sum of the numbers in any row, column, or diagonal is 3, the sum of the numbers in the shaded squares above is $3 \times 3 = 9$.

Notice that the 9 shaded squares above include all six squares in the left and right columns, plus three copies of the center square, as shown below:

 (3 copies of the center number)

So, the sum of the numbers in the left and right columns plus 3 copies of the center number is 9.

The sum of the numbers in any column is 3, so the sum of the numbers in the left and right columns is $3 \times 2 = 6$.

Therefore, the sum of three copies of the number in the center square is $9-6=3$. So, the center number is $3 \div 3 = 1$.

Next, we look at the remaining squares. We see that a number in a corner is part of three different sums, while the remaining non-corner numbers are only part of two different sums.

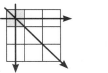

The only two groups of three integers with a sum of 3 and that contain -3 are (-3, 1, 5) and (-3, 2, 4).

So, -3 cannot be placed in a corner.

We choose one of the four non-corner squares to place -3 in.

	1	-3

$\boxed{5}+1+(-3)=3.$ So, to complete this row, the number in the left square of the row is 5.

5	1	-3

The only other sum that includes -3 is $-3+2+4$. So, we choose a corner to place the 2 in and place the 4 to complete this line of three numbers.

		2
5	1	-3
		4

We use the sums of the rows, columns, and diagonals to find the placements of the remaining numbers:

-2		2		-2		2		-2	**3**	2
5	1	-3		5	1	-3		5	1	-3
		4			**0**	4		0	**-1**	4

We could have placed -3 in any of the four remaining non-corner spots, and we could have switched the place of the 2 and 4 around the -3.

The 8 valid arrangements of these numbers into a Magic Square are shown below.

-2	3	2		-2	5	0		0	-1	4		0	5	-2
5	1	-3		3	1	-1		5	1	-3		-1	1	3
0	-1	4		2	-3	4		-2	3	2		4	-3	2

2	-3	4		2	3	-2		4	-3	2		4	-1	0
3	1	-1		-3	1	5		-1	1	3		-3	1	5
-2	5	0		4	-1	0		0	5	-2		2	3	-2

For additional books, printables, and more, visit
BeastAcademy.com